Urban Sustainability

The Urban Sustainability Book Series is a valuable resource for sustainability and urban-related education and research. It offers an inter-disciplinary platform covering all four areas of practice, policy, education, research, and their nexus. The publications in this series are related to critical areas of sustainability, urban studies, planning, and urban geography.

This book series aims to put together cutting-edge research findings linked to the overarching field of urban sustainability. The scope and nature of the topic are broad and interdisciplinary and bring together various associated disciplines from sustainable development, environmental sciences, urbanism, etc. With many advanced research findings in the field, there is a need to put together various discussions and contributions on specific sustainability fields, covering a good range of topics on sustainable development, sustainable urbanism, and urban sustainability. Despite the broad range of issues, we note the importance of practical and policy-oriented directions, extending the literature and directions and pathways towards achieving urban sustainability.

The series will appeal to urbanists, geographers, planners, engineers, architects, governmental authorities, policymakers, researchers of all levels, and to all of those interested in a wide-ranging overview of urban sustainability and its associated fields. The series includes monographs and edited volumes, covering a range of topics under the urban sustainability topic, which can also be used for teaching materials.

Xiaoyu Lin · Han Wang

A Multi-Level Analysis of China's Rural Built Environment

Macro-Spatial Governance in Xiaqiao Village (Volume I)

Springer

Xiaoyu Lin
School of Architecture
Harbin Institute of Technology
Shenzhen, Guangdong, China

Han Wang
College of Fashion and Design
Donghua University
Shanghai, China

ISSN 2731-6483 ISSN 2731-6491 (electronic)
Urban Sustainability
ISBN 978-981-95-4944-3 ISBN 978-981-95-4945-0 (eBook)
https://doi.org/10.1007/978-981-95-4945-0

This book was funded by the National Natural Science Foundation of China (No. 52208018).

This Springer imprint is published by the registered company Springer Nature Singapore Pte Ltd.
The registered company address is: 152 Beach Road, #21-01/04 Gateway East, Singapore 189721, Singapore

Preface to the Series

The transformation of rural China represents a significant socio-spatial phenomenon. While urbanization has received extensive scholarly attention, the dynamics of change within rural built environments require further investigation. These areas are not static backgrounds but evolving systems influenced by socio-economic, political, and cultural shifts. This series addresses this subject by presenting a comprehensive analysis of the rural built environment, examining it as a dynamic system shaped by the interaction of time, space, and human agency.

The central premise of this research is that rural transformations follow autonomous trajectories, determined by the interplay between institutional mandates and grassroots agency across multiple scales. To investigate this complexity, we established a seven-level analytical framework. This methodology integrates concepts from urban morphology and Open Building theory to examine the built environment across a spectrum of scales: from the macro-level (Region, Settlement, Sub-settlement) to the micro-level (Neighborhood, Building Cluster, Individual Structure, and Infill). This framework allows for the reconstruction of spatial configurations and architectural forms across four distinct socio-economic periods: the Qing Dynasty, Republican China, the planned-economy period, and the market-economy period (1980-2015).

The research relies on an in-depth case study of Xiaqiao Village in Taishun County, Zhejiang Province. With a history dating back to the Five Dynasties period and a substantial number of architectural remains still in use, Xiaqiao provides a stable empirical foundation. The study utilizes field surveys, historical archives, family genealogies, and interviews to reconstruct the village's architectural and spatial evolution from 1644 to 2015.

Given the complexity of the seven-level framework, this research is organized into two interdependent volumes. This division is based on the hierarchy of control and the nature of the governing agents.

The seven levels form a nested system where each higher level serves as the physical framework for the level below it. However, a significant shift in governance occurs between the sub-settlement (Level 3) and the neighborhood (Level 4).

Volume I (This Volume): ***Macro-Spatial Transformations and Governance*** focuses on the first three levels: **Region**, **Settlement**, **and Sub-settlement**. These levels constitute the collective environment. They are primarily shaped by "top-down" forces, including natural geography, lineage organizations, and formal government policies. In this context, the settlement serves as the structural framework within which community life is organized.

Volume II: ***Micro-Architectural Adaptation and Agency*** focuses on the subsequent four levels: **Neighborhood**, **Building Cluster**, **Individual Structure**, **and Infill**. These levels constitute the domain of domestic life. They are primarily shaped by "bottom-up" forces, driven by individual families and residents. Here, the focus shifts to how individuals act as agents to adapt buildings and interiors within the constraints established by the macro-spatial framework.

A primary contribution of this series is the exploration of the governance hierarchy underlying spatial transformations. By dissecting the roles of various agents—including government bodies, intermediate organizations (such as lineage groups), and individual households—the research reveals how control shifts across different spatial levels. Government power tends to exert greater influence at higher spatial levels (discussed primarily in Volume I), while individual agency determines outcomes at the building and interior levels (discussed primarily in Volume II).

This series offers data and theoretical perspectives relevant to academic discourse and practical application. First, it provides a refined methodology for rural studies through the seven-level framework. Second, it offers empirical data on indigenous vernacular architecture by investigating common buildings rather than solely focusing on heritage sites. Third, it clarifies the relationship between governance structures and physical space.

The conceptual framework and findings presented in this series offer transferable principles for rural development and cultural conservation. It serves as a reference for scholars in urban studies, architectural history, sociology, and planning, as well as for policymakers involved in rural revitalization.

Shenzhen, China Xiaoyu Lin

Acknowledgements This book series is the culmination of years of dedicated research, tracing its origins back to Xiaoyu's doctoral studies at The University of Hong Kong (HKU) from 2013 to 2018. The comprehensive analysis presented herein is built upon an intensive, five-year ethnographic and architectural fieldwork in traditional rural areas across China, with a particular focus on the distinctive cultural landscapes and built environments of southern Zhejiang Province. During this period, Xiaoyu was privileged to collect an extensive archive of firsthand architectural survey data, alongside rich qualitative insights, which form the empirical backbone of this scholarly endeavor.

The deepest gratitude is owed to our supervisor, Professor JIA Beisi, whose unparalleled guidance, intellectual rigor, and unwavering support were instrumental at every stage of this journey. His profound insights, meticulous feedback, and boundless encouragement transformed this research from a nascent idea into a structured and impactful study. We are truly fortunate to have benefited from his mentorship.

We would also like to express our sincere appreciation to the faculty and administrative staff of the Faculty of Architecture at the University of Hong Kong. Their efficient assistance and supportive environment greatly facilitated my research and doctoral experience. The exceptional resources provided by the University Libraries, including their vast collection of academic literature and digital databases, were invaluable in shaping the theoretical and contextual foundations of this work.

This academic journey was profoundly enriched by the camaraderie and intellectual exchange with our fellow postgraduate students. We extend heartfelt thanks to Dr. Dinghang Wu, Dr. Yiqi Tao, Dr. Wenjian Pan, for the stimulating discussions on rural morphology, and to Dr. Zhu Xu, Dr. Liran Chen for their insights into historical archives. Their companionship, critiques, and shared experiences created a vibrant and supportive academic community that made the challenges of doctoral research far more manageable and enjoyable.

Finally, and most profoundly, we dedicate this work to our families. Our deepest gratitude goes to our parents, for their unconditional love, enduring encouragement, and steadfast financial and emotional support throughout our academic career and life. Their belief in our pursuits provided the essential foundation for us to embark on and complete this ambitious project.

This book would not have been possible without the collective support, intellectual contributions, and personal sacrifices of so many individuals. We are immensely grateful to all of them.

Competing Interests The authors declare that they have no known competing financial interests or non-financial interests in relation to the content of the book.

Contents

About the Authors

Xiaoyu Lin is an Associate Professor at the School of Architecture, Harbin Institute of Technology (Shenzhen). She received her Master of Engineering in Architectural History and Theory from Southeast University in 2013 and her Ph.D. in Philosophy from the University of Hong Kong in 2018. Her research focuses on traditional regional architectural culture and theory, including theories and practices of urban and rural renewal, traditional rural spatial landscapes, adaptive design methods for traditional spaces, and architectural anthropology.

Han Wang is an Associate Professor in the Department of Environmental Design at Donghua University. She received her Bachelor of Architecture from Dalian University of Technology and his Master of Architecture from the University of Melbourne. In 2018, she earned her Ph.D. in Philosophy from the University of Hong Kong. Her research focuses on urban spatial morphology, including the renewal and revitalization of urban public spaces, the application of morphological methods in historical residential areas, and participatory design and community renewal studies.

List of Figures

List of Tables

Chapter 1
Introduction

1.1 Rural Regeneration in Contemporary China

The vast rural areas of China have undergone a series of comprehensive transformations over the past few centuries, encompassing both social and spatial dimensions. Transitioning from a premodern society prior to the late nineteenth century, the nation has gradually evolved into a modern society by the twenty-first century. These dramatic shifts in social structures have precipitated corresponding changes in the built environment. Heated debates over urbanization have persisted in China since the 1980s. According to data from six national population censuses conducted in 1953, 1964, 1982, 1990, 2000, and 2010, China's urbanization rates were recorded at 12.84%, 17.58%, 20.43%, 25.84%, 35.39%, and 49.68% respectively. By 2011 and 2014, these figures rose to 51.27% and 54.77%. Over the past three decades, China has experienced peak urbanization.

However, this rapid urbanization has resulted in imbalanced social development characterized by an urban-rural dualistic structure. Rural communities have assumed a vulnerable position within one of history's largest urbanization processes. They have been swept up in the powerful currents of modernization while remaining marginalized in both academic discourse and practical planning initiatives.

1. Transformation of Social Structure

   Chinese rural society has undergone a profound transformation from traditional to modern systems over the past century. This shift manifests across multiple dimensions: the gradual decline of the kinship-based family system, the rising prominence of individual agency, and the displacement of collective living patterns by modern individualized lifestyles. In traditional agrarian communities, clans and extended families functioned as fundamental socio-economic units, playing pivotal roles in governance, resource distribution, and cultural transmission. However, modernization processes have eroded these entrenched

X. Lin and H. Wang, *A Multi-Level Analysis of China's Rural Built Environment*, Urban Sustainability, https://doi.org/10.1007/978-981-95-4945-0_1

structures, concurrently elevating the societal status of individuals. Such structural metamorphosis has not only reconfigured rural social organization but also exerted transformative influences on the built environment.

2. Reshaping of Spatial Environments

   The rural built environment has experienced profound transformations amidst rapid urbanization. Accelerated urban migration has precipitated dramatic demographic shifts and functional reconfigurations in rural areas. Traditional architectural forms—including historic villages, vernacular dwellings, and communal structures—face existential threats from demolition and haphazard redevelopment. Contemporary rural constructions frequently employ generic designs and industrialized materials, creating stark visual and cultural dissonance with traditional landscapes. This spatial restructuring has engendered dual consequences: while physically altering rural morphologies, it has also undermined cultural continuity and weakened community cohesion.

3. Built Environment as a Social Structure Manifestation

   The rural built environment serves as a tangible embodiment of evolving social structures. Spatial configurations mirror not only organizational modes of rural life but also shifting relational dynamics within communities. The evolution of public spaces exemplifies this correlation: traditional civic nodes like ancestral halls, village temples, and communal squares historically functioned as social nexuses, facilitating collective interaction and mutual aid. Their contemporary marginalization and replacement by privatized spaces epitomize the transition from collectivist to individualist social paradigms—a transition concomitantly reflected in diminishing community solidarity.

4. Architectural Heritage as Cultural Repository

   Vernacular architectural elements constitute vital repositories of localized cultural knowledge. These material artifacts encode historical narratives while sustaining collective memory and place-based identity. Preserving rural built heritage holds critical significance for safeguarding cultural diversity, as traditional settlements—with their distinctive stylistic vocabularies and embedded socio-cultural codes—provide invaluable resources for historical research. Nevertheless, modernization-driven demolition and the proliferation of placeless architecture are precipitating irreversible cultural erosion. Systematic study of rural architectural transformations therefore serves dual purposes: enhancing comprehension of socio-historical processes while informing evidence-based cultural preservation strategies.

This research is situated within the context of transformative changes sweeping across rural China, where a new round of "New Village" construction planning initiatives is currently underway. At this critical juncture, it becomes imperative not only to conduct a retrospective analysis of rural societal evolution but also to re-examine historical trajectories in the transformation of rural built environments.

The study establishes theoretical frameworks for strategic interventions that balance heritage conservation with sustainable development. It provides actionable guidelines not only for architectural rehabilitation and new construction projects, but also proposes an innovative paradigm for understanding rural spatial transitions. These contributions aim to inform policymaking and enhance operational effectiveness in China's rural revitalization practices.

1.2 Research Content

This research explores the transformation process of the rural built environment in China. It tries to establish a dynamic view to study the space comprehensively that contains multiple dimensions (time, space, people), multiple levels of space, and multiple periods.

Generally, this research refers to one case, two systems, three agents, four periods, seven levels and multi-factors

- One Case: this research is mainly an in-depth case study of the Xiaqiao Village in Zhejiang Province, China.
- Two Systems: self-govern system 'bottom-up'; governmental system 'top-down.'
- Three Agents: government, villagers (family and individual), and intermediate organizations
- Four Socio-economic Periods: the imperial Qing dynasty (1616–1911), the Republican China (1912–1948), the planned-economy period (1949–1979), and the market-economy period (1980–2015).
- Seven Levels: three levels of settlement that are region, settlement, sub-settlement; four levels of building that are neighborhood, building group, building enveloping and structure and space, infill, and furnishing and equipment
- Multiple Factors: society (economic, political, technological, cultural, religious, etcetera), everyday life (household, family structure, living mode, production method, etcetera)

1.3 Research Case: A General Profile the Xiaqiao Village

The research uses a specific village, the Xiaqiao Village, to conduct an in-depth case study of the transformation of the rural built environment. Xiaqiao is a typical village located among the mountains in the south of Zhejiang province. It belongs to the Sixi (泗溪) township, Taishun (泰顺) county, Wenzhou city, along with the administrative hierarchy (Fig. 1.1).

It is beneficial to define the small village within a larger background of the entire Taishun County first.

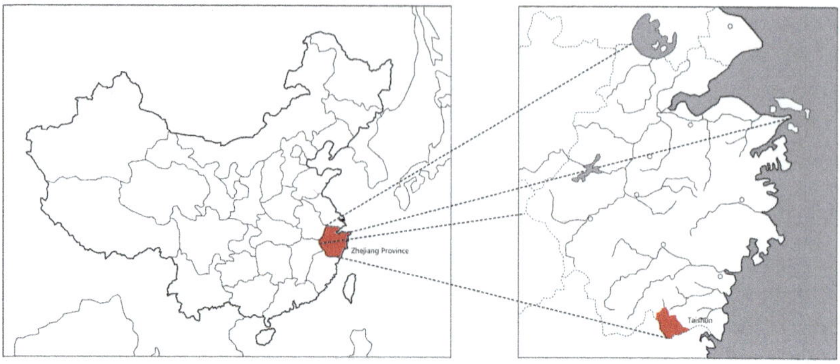

Fig. 1.1 Location of Taishun

1.3.1 A General Profile of the Taishun County

1.3.1.1 Define Taishun on Space

A county is the fourth-level administrative division in China. The country is divided into thirty-four regions at the second level. In this case, it is Zhejiang province. Each region has several districts at the third level; in this case, it is Wenzhou city. Taishun belongs to the fourth level and is one of the eleven secondary administrative districts under Wenzhou.

Taishun County is subdivided into ten townships and eleven resident districts, each of which consists of several villages (Fig. 1.2). The number of villages is not the same for each township and resident district, and they differ in size and population.

From an administrative neighborhood perspective, Taishun lies on the boundary between Zhejiang and Fujian provinces. Among the seven other counties surrounding it, three belong to Zhejiang province on its north and east sides, namely Cangnan (苍南), Jingning (景宁), and Wencheng (文成); four belong to Fujian province

Fig. 1.2 Township inside Taishun, (Source: redraw based on [2])

on its south and west sides, namely Fuding (福鼎), Zherong (柘荣), Fu'an (福安), and Shouning (寿宁). The role as a juncture between Zhejiang and Fujian provinces makes research on Taishun more meaningful from the perspective of regional communication.

From the perspective of geographic research, Taishun lies among the coastal mountainous area in southeastern China. *The coastal mountainous area in southeastern China* (afterwards CMASC, 东南沿海丘陵地带) is both a concept of natural geography and cultural geography, which contains the mountains along the southeast coastal line in China consisting of South Zhejiang, East Guangdong and the whole area of Fujian province [1]. This area is connected together by a series of mountains with an average elevation between 1000 and 1500 meters. They also share a similar geographical environment and cultural features. Unlike the great plain area with convenient traffic conditions, the CMASC where Taishun is located is divided into a great number of independent basins by mountains and rivers, nurturing several sub-culture systems correspondingly (Fig. 1.3), while sharing common features generally. Thus, the research of Taishun, as a typical case study, could be counted as a miniature of the whole area to some extent.

In history, Taishun has a close relationship with its adjacent regions. Indeed, more areas should be primarily investigated as references based on the extent of their communication with Taishun. The potential areas include Wenzhou city (温州市),

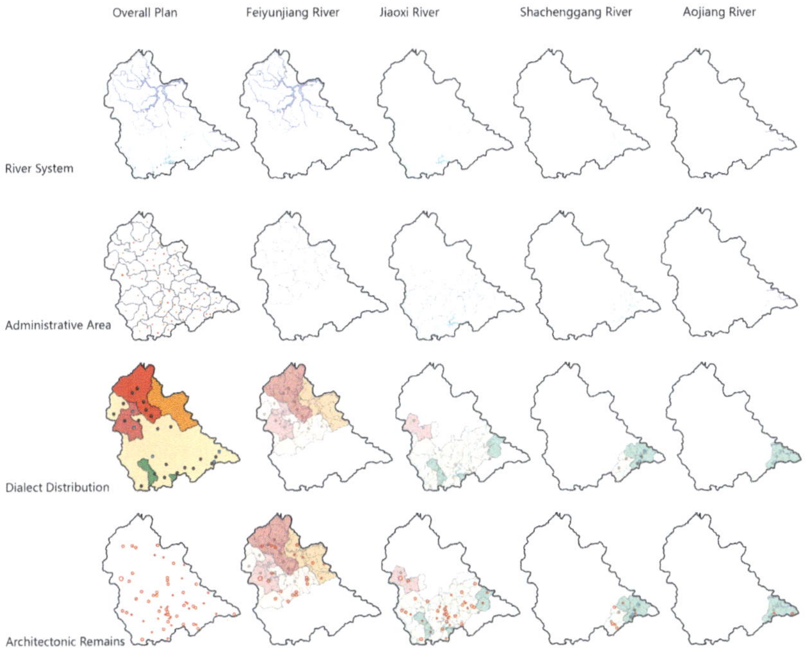

Fig. 1.3 The Relationship between Architectonic Remains, Geographical Environment, Administrative Districts, Dialect District in Taishun, (Source: redraw based on [2])

Lishui (丽水), Rui'an (瑞安), Fu'an (福安), Fuding (福鼎), Wencheng county (文成县), Cangnan county (苍南县), Pingyang county (平阳县), Qingtian county (青田县), Yunhe County (云和县), Shouning county (寿宁县), Qingyuan county (庆元县) etc. In order to explore the impact of immigration, their original hometowns can be involved in the research as well, such as Tong'an (同安) in the south of Fujian province.

1.3.1.2 Define Taishun on Time

The major historical period concerned in my research is from 1644 to 1912, the Qing Dynasty. The year 1644 is the time when the army of the Qing occupied Beijing and the Qing emperor officially became the ruler of China. Thus, in this research, the year 1644 is used as the beginning of the period. But considering the continuity of crafts and techniques, several important architectural remains built before or after this designated period (1644–1912) would be involved as references. Such examples include the Shiying Gate (Shiyingmen, 士英门), Luoyang Stone Pavilion (Luoyang Shiting, 罗阳石亭), built in the Ming Dynasty, or several big houses built or rebuilt during the Republic of China.

Historically, Taishun County was officially established by the court of the Ming Dynasty in the third year of Jingtai (1452), integrating several areas together that were separated from Yixiang Township (Yixiang xiang, 义翔乡) of Rui'an City and Guiren Township (Guiren xiang, 归仁乡) of Pingyang County. Although people had already lived in Taishun for a long time before 1452, including immigration of the Han nationality (汉族) from central China (中原) since the Tang Dynasty and the native She minorities (畲族) whose ancestors belonged to the ancient Yue nationality (古越人), the formation of a local government implied growing attention from the court, and a large-scale regional development was about to start.

The administrative adjustment here was caused by rampant banditry happened around 1452. It was a revolt by silver miners who occupied the mountainous area around the boundary of Zhejiang and Fujian provinces against the government's prohibition of private mining. The armed rebellion spread among the mountains and devastated most buildings there. In 1452, after the rebellion was suppressed by the government, several new counties, including Taishun, were established successively by reorganizing the lands originally belonging to other counties, in order to strengthen control over the previously uncontrolled area hidden among remote mountains. Literally, Taishun means peace, allegiance, and obedience, which reveals the court's resolution to ensure powerful governance here. From then on, until the end of the Qing Dynasty, several other rebellions spread to Taishun: in 1559, 1636, 1644, 1649, and 1674 by pirates from Fujian province; in 1648 and 1657 by armed volunteers upholding the Ming Dynasty against the Qing Dynasty; in 1858 by the remnants of the Taiping Rebellion (太平天国). Most of the turmoil occurred during the Ming Dynasty and the early Qing Dynasty, except for one instance caused by the Taiping Rebellion in the late Qing Dynasty. Their effects on Taishun, especially on the remaining situation of the architecture, need to be further ascertained.

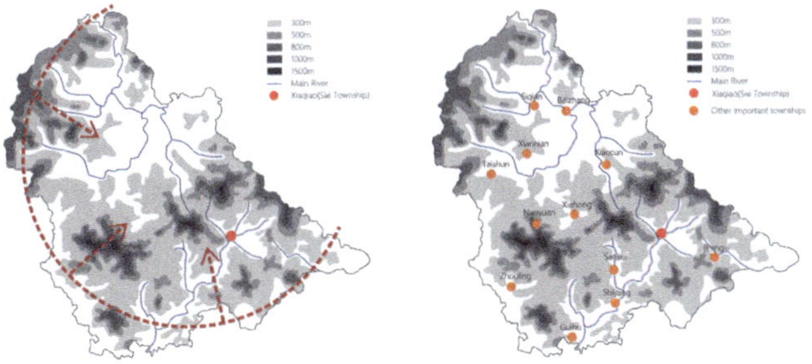

Fig. 1.4 Geography around Taishun, (Source: redraw based on [2])

1.3.1.3 Define Taishun on Culture

Cultural Transition Region

Between Min and Zhe Cultural Zones Taishun is located in the transitional area between the Min Sub-cultural Zone in the Southeast Cultural Zone (SCZ) and the Zhe Sub-cultural Zone in the Yangzi Cultural Zone (YCZ) [1, 3, 4]. Unlike other areas in Zhejiang Province, the culture in the area where Taishun lies incorporates significant influences from North Fujian, even South Fujian, constantly brought by daily trade or sporadically brought by immigration. It can be demonstrated by Taishun's dialect system, which is a hybrid of Wu dialect (dominant in other areas of Zhejiang Province) and Mindong dialect (dominant in eastern Fujian Province) [5]. Among them, the most popular one is recognized as a Fujian dialect called Manjiang (蛮讲). Hereby, Taishun can be understood as a mixture generated by the two cultures dominated by Fujian influences.

Between Higher Mountains and Lower Coastal Plain From the perspective of topography, Taishun is located in the secondary coastal mountains, sandwiched between higher mountains in the hinterland, like Lishui city, and lower coastal plains, like Ruian city and Cangnan county (Fig. 1.4). This makes it a traffic junction between the mountains and the sea. Whether it is outside commodities from the sea being transported inward to higher mountainous areas or raw materials being transported outward to the port, their transport cannot avoid passing through Taishun by road [2]. The role as a connected station between the sea and the higher mountains makes it possess both the qualities of an extroverted marine culture and an introverted mountainous culture. For example, the common dwellings in Taishun look simple like those in other mountainous regions, but the ancestral halls are gorgeous and complicated, with delicate details, similar to those seen in coastal areas (Fig. 1.5).

Fig. 1.5 Left: Common dwelling in Daxiyuan Village of Taishun; Right: Ancestral hall of Family Bao in Yuyan Village of Taishun

Internal Cultural Complexity and External Cultural Hysteresis

The complex geographic environment has endowed Taishun with a self-contained cultural system, yet its preservation of ancient customs reveals a culturally hysteretic character compared to other regions.

Complex Culture Network Inside Taishun is situated in southeastern mountainous terrain, characterized by independent drainage systems formed by complicated topography. Four major rivers and their tributaries dominate the region: the upper Feiyunjiang River (north), Aojiang River, Jiaoxi River, and Shangchenggang River (south) (Fig. 1.3). Each river basin is further segmented by high mountains and deep valleys, fostering independent development of sub-regional cultures. This has created China's most intricate cultural landscape among the Han ethnicity, exemplified by its complex dialect system.

Cultural Hysteresis Compared to Other Regions Taishun's geographical isolation, created by surrounding mountains, has limited its interaction with external regions. During the agricultural era, when transport relied primarily on human and animal labor, the mountainous terrain made travel to other regions extremely difficult. This isolation has preserved local traditions intact, resulting in a construction system that retains archaic techniques and styles.

A case in point is the Woven Timber Arch-Beam Bridge (编木结构拱桥; Fig. 1.6), a form depicted in the Northern Song Dynasty painting Along the River During the Qingming Festival (清明上河图). This bridge type, still constructed in Taishun and adjacent counties, exemplifies the coexistence of advanced timber engineering and traditional craftsmanship. Although bridges fall outside this study's scope, their sophisticated structures demonstrate how Taishun maintains a cultural continuity that contrasts with regions undergoing rapid modernization.

Fig. 1.6 Wenxing pavilion bridge in Xiaocun Village of Taishun

The Effect of Immigration

Another key factor that impacts Taishun's culture is immigration. In history, Taishun has a large area but a small population, which attracts people from other places to move in. Thus, the immigrants who came successively became a dominant power engaging in local development [2].

Firstly, Taishun was influenced by the diffusion of Central China Culture, a common feature in the Southeast Culture Zone [3]. During several waves of southward migration in premodern China, ancient Han culture was introduced to the region. This Han culture blended with the indigenous Yue culture, forming a hybrid local culture that retains elements of both traditions.

Secondly, most current residents are descendants of migrants from Zhejiang and Fujian who arrived during the Ming and Qing dynasties, bringing distinct cultural practices that remain visible today. For example, local beliefs vary according to the migrants' regions of origin. By tracing these patterns, the pathways of cultural transmission through immigration can be analyzed. Thus, immigration has played a crucial role in fostering Taishun's culturally diverse hybridity.

Overall, Taishun's cultural hybridity—resulting from the integration of diverse regional and historical influences—makes it a valuable case study for understanding cultural transmission and adaptation.

1.3.2 A General Profile of the Xiaqiao Village

Xiaqiao Village, situated in the southwestern sector of Sixi Town, Taishun County, Zhejiang Province, is geographically defined by its unique hydrological configuration where four streams converge—a characteristic encapsulated in the local toponym "Sixi Hui Lan" (四溪汇澜, Four Streams Convergence). This mountainous settlement is renowned for its ecological landscape and cultural heritage, most notably housing the Sisters' Bridge (姊妹桥), a nationally protected timber corridor bridge acclaimed as "China's most architecturally significant covered bridge." As of the 2020 census, the village comprises 744 households with a registered population of

2331 residents, spanning a total area of 9716 acres (39.3 km^2), including 648 acres (2.62 km^2) of arable land and 6054 acres (24.5 km^2) of forested terrain.

With a history dating back to the Five Dynasties period (907–960 CE), Xiaqiao Village has witnessed significant transformations in its natural and built environments over the past millennium. Today, its population primarily descends from four lineages: Chen, Tang, and two branches of Lin, whose ancestors migrated to the village during different historical periods. This study analyzes the village's autonomous self-housing system from the late Ming Dynasty to the present.

In premodern China, Xiaqiao was a self-governing village dominated by lineage-based systems. During the Qing Dynasty, the four lineages established a co-governance model that shaped local power dynamics through competition and cooperation, profoundly influencing the village's spatial development. The transition to modernity brought profound societal changes: local governance shifted to centralized administrative bodies, the economy diversified beyond agriculture, and family structures transformed from extended households to nuclear families of three to five members. These shifts drove radical transformations in both the built environment and architectural forms.

A great number of precious architectural remains built in the Qing Dynasty are kept in Xiaqiao Village as well as the whole Taishun County. The abundant sources of existing architectural resources presage a bright prospect for thoroughly conducting in-depth research on the rural built environment.

The architectural remains cover different types and construction periods. Folk houses with well-decorated timber structures remain there, including several architectural types like dwellings, ancestral halls, temples, public pavilions, etc. They were built from the end of the Ming Dynasty to the present.

In contrast to heritage sites conserved by a museum-like approach, most vernacular buildings surviving in Xiaqiao Village are still in use. They still serve their original functions. They offer not only the convenience of conducting surveys by directly entering the buildings, but also a precious opportunity to observe how architecture and people interact. These buildings facilitate both in-situ surveys and the synthesis of the transformation process. The extensive architectural vestiges spanning multiple historical epochs constitute a valuable research specimen for deciphering the evolutionary logic of rural spatial morphology.

1.4 Book Structure

This research is divided into two volumes to cover the seven-level analytical framework. Volume I focuses on the macro-spatial levels (Region, Settlement, and Sub-settlement), analyzing the physical and institutional context of the village. Volume II focuses on the micro-spatial levels (Neighborhood, Building Cluster, Individual Structure, and Infill), analyzing architectural adaptation and daily usage.

Volume I: Macro-Spatial Transformations and Governance consists of six chapters:

Chapter 1 outlines the research background, the context of rural regeneration in China, and the organization of the two volumes.

Chapter 2 presents the theoretical framework. It defines the research scope regarding the rural built environment and states the primary objectives. This chapter connects the case study of Xiaqiao Village to broader theories of rural planning and development.

Chapter 3 describes the research methodology. It defines the complete seven-level analytical framework derived from urban morphology and Open Building theory. Although this volume applies only to the first three levels, this chapter explains the full system to provide a complete methodological context.

Chapter 4 analyzes the macro-spatial levels: Level 1 (Region), Level 2 (Settlement), and Level 3 (Sub-settlement). Using historical archives and spatial mapping, this chapter tracks the geographic and social changes of the village. It examines the relationship between spatial patterns, population changes, and administrative structures across four historical periods.

Chapter 5 examines governance at the settlement levels. It identifies the agents that regulate land use, public space, and settlement morphology, such as government bodies and lineage organizations. The chapter analyzes how these agents exert control and how their influence has changed over time.

Chapter 6 provides the conclusion for Volume I. It summarizes the findings regarding macro-spatial changes and governance. This chapter outlines the external constraints and institutional contexts that influence the village, serving as a basis for the micro-level analysis in Volume II.

References

1. Skinner G, Henderson Mark, Yuan Jianhua. (2000). China's Fertility Transition through Regional Space. Social Science History. 24:613–652. doi:10.1017/S0145553200010312
2. Wu, Songdi 吴松弟, Liu Jie 刘杰. (2009). Zouru Zhongguo de Chuantong Nongcun 走入中国的传统农村:浙江泰顺历史文化的国际考察与研究(An International Exploration and Research on the Historical Culture of Taishun, Zhejiang: Stepping into Traditional Rural China). Jinan: Qilu Publish. pp 38-39.
3. Wu Bihu 吴必虎. (1996). "Zhongguo Wenhuaqu de Xingcheng yu Huafen" 中国文化区的形成与划分 (The Formation and Division of Cultural Regions in China). *Academic Monthly* 03:10–15.
4. Wu Songdi 吴松弟. (2010). Zhongguo Dongnan Shanqu de Diyu Shehui Jiegou: Yi Ming-Qing Zhejiang Taishunxian Weili 中国东南山区的地域社会结构:以明清浙江泰顺县为例 (Regional Social Structure in the Mountainous Southeast of China: A Case Study of Taishun County, Zhejiang, in the Ming and Qing Periods). Beijing: Commercial Press.
5. Zhu Haibin 朱海滨. (2010). Jinshi Zhejiang Wenhua Dili Yanjiu 近世浙江文化地理研究 (Research on the Cultural Geography of Zhejiang in the Early Modern Period). Shanghai: Fudan University Press

Chapter 2
Theoretical Framework

2.1 The Complexity of Rural Built Environment in China

The rural built environment represents a distinct category parallel to its urban counterpart, yet exhibits unique developmental characteristics. Unlike urban environments, which function as unified entities capable of fulfilling all daily necessities, the rural built environment operates as an interconnected network of semi-autonomous settlements. Each village functions as an independent living unit within vast rural territories, addressing fundamental needs through basic housing, agricultural production, and communal social structures. These settlements collectively form cohesive social networks through functional interdependencies. Historically, rural villages maintained stable operational continuity as foundational human habitats throughout China's agrarian era.

China's settlement patterns remain predominantly rural in scale and significance, with approximately 53.4% of its 1.4 billion population (713 million people) residing in over 3.8 million villages that collectively contain more than 300 million square meters of built infrastructure. Contemporary modernization efforts are radically transforming these environments under dual pressures of urbanization and industrialization. This restructuring process manifests most visibly through dynamic spatial reorganizations and emerging cluster-based development patterns across rural landscapes.

Chinese rural settlements demonstrate exceptional resilience through centuries of continuous occupation while urban areas have proven vulnerable to sociopolitical upheavals. Rural communities exhibit remarkable capacity to preserve cultural traditions across shifting governance systems—a continuity physically manifest in enduring architectural forms and spatial configurations. Today's rural landscapes present a palimpsest of temporal layers: modern terraced housing complexes stand adjacent to centuries-old timber-frame dwellings, ancestral kinship systems continue operating alongside contemporary municipal governance structures in

X. Lin and H. Wang, *A Multi-Level Analysis of China's Rural Built Environment*, Urban Sustainability, https://doi.org/10.1007/978-981-95-4945-0_2

village administration, and traditional ritual spaces coexist with modern public facilities.

Rural settlements not only physically endure over vast timescales but also persist through epochs marked by drastically different social, economic, and political contexts. Typically, a rural habitat established before the Qing dynasty would traverse four distinct periods: the premodern era (pre-1912), the Republican era (1912–1949), the planned-economy period (1949–1978), and the market-economy period (1978–2015). Architectural remnants and spatial configurations from these periods bear imprints of their socio-economic characteristics, forming a palimpsest of historical layers. Even settlements sharing identical geographic locations, scales, and demographic compositions can evolve into remarkably distinct communities due to their unique developmental trajectories or divergent social origins of their founders.

While rural settlements may appear to serve relatively singular societal roles, their spatial diversity manifests across multiple dimensions. They manifest in varied scales, social hierarchies, and functional typologies, accommodating diverse human activities. Beyond fundamental spatial functions—such as residential dwellings, lineage ancestral halls, and local temples—recent decades have seen the introduction of novel spatial forms resulting from hybridized functional demands.

Moreover, the human dimension profoundly shapes rural built environments, as these spaces are constructed by, for, and through human agency. This triadic relationship introduces a critical layer of complexity: governmental authorities increasingly regulate rural spatial planning with heightened specificity; individual residents primarily focus on constructing/managing their own dwellings with limited concern for collective interests; and lineage groups—as unique social actors—leverage communal bonds to actively engage in local governance, advocating for the interests of their constituent members through coordinated efforts.

Recognizing built environments as dynamic constructs shaped by human action necessitates acknowledging their reciprocal influence on human behavior. Spatial forms not only reflect societal roles but also actively shape them—long-term control over physical environments inevitably induces behavioral adaptations.

Given the ceaseless transformations across rural China's vast territories over centuries, contemporary rural landscapes emerge as true kaleidoscopes. They juxtapose historical and modern elements, vernacular traditions and global influences, manifesting in layered spatial complexes. Here, diverse architectural typologies coexist—from traditional timber-frame dwellings and ancestral halls to modern concrete structures, alongside temples, schools, commercial spaces, and civic buildings. These spatially layered entities, imprinted with the traces of social power struggles across different historical epochs, ultimately coalesce into an intricate tapestry of spatiotemporal interplay.

2.2 The Issues and Objectives

2.2.1 *Main Proposition*

The rural built environment constitutes a dynamic system integrating multidimensional temporal, spatial, and people. Perpetual flux rather than stasis defines these landscapes. Transformations emerge not from singular political or economic drivers, but through the confluence of institutional mandates and grassroots agency. These forces exhibit inherent mutability, exerting multi-scalar influences that reshape built environments with varying intensity and scope.

Conventional scholarship often frames rural transformations as mere byproducts of urban-centric urbanization. While these processes interact dialectically, they constitute distinct phenomena governed by divergent logics, temporalities, and socio-structural frameworks. Rural spatial metamorphoses follow autonomous trajectories irreducible to urban development paradigms.

This research seeks to elucidate the dynamic transformation processes within China's rural built environment through a detailed case study of Xiaqiao Village (下桥, under bridge). It aims to reconstruct the village's architectural and spatial evolution across four centuries (1644–2015), spanning four socio-economic periods: the Qing dynasty (1644–1911), Republican China (1912–1948), the planned-economy period (1949–1978), and the market-economy period (1979–2015). The study investigates how shifting societal conditions trigger cascading architectural transformations, revealing the intricate relationships between macro-level social changes and micro-level spatial adaptations.

2.2.2 *Shifting Research View*

Rural spaces remain conspicuously absent from mainstream architectural discourse. Prevailing discussions on the built environment predominantly focus on urban phenomena—metropolitan expansion, capitalist urbanization, and the commodification of public space. The rural built environment occupies a distinct conceptual territory. Unlike monumental urban architectures designed as symbolic landmarks, rural structures exist as lived environments intrinsically tied to daily realities. Compared to formally rigid urban typologies, these vernacular landscapes possess inherent adaptive capacities to evolve with communal life patterns.

Despite this potential, architectural scholarship in China continues to prioritize elitist conceptions of "nobility," systematically marginalizing everyday spatial practices. Rural communities—constituting the foundational stratum of Chinese society—have endured prolonged neglect within modernization narratives. Consequently, rural studies emerge as a critical lens for interrogating, deconstructing, and reimagining quotidian spatial experiences. While rural landscapes offer unparalleled opportunities to engage with lived realities, substantive discourse in this domain remains acutely underdeveloped.

2.2.2.1 An Integral View to Research on Rural Built Environment

Existing scholarship on rural built environments predominantly bifurcates into settlement studies and architectural typology analyses. Such disciplinary segregation artificially isolates structures from their contextual ecosystems, neglecting critical examination of their reciprocal influences. This fragmented approach reveals a fundamental deficiency in holistic comprehension of rural spatial systems.

Settlements and buildings inherently constitute complementary scales of environmental analysis. While employing distinct observational frameworks—macro-scale spatial patterns versus micro-level material configurations—they collectively describe unified socio-spatial phenomena. This ontological interdependence necessitates integrated methodologies that bridge scalar perspectives.

Current research disproportionately emphasizes specialized architectural taxonomies rather than developing comprehensive theoretical models for rural contexts. This study proposes an alternative analytical paradigm, advancing synthetic frameworks to reinterpret China's rural transformations. Through multi-scalar synthesis, it seeks to rejuvenate theoretical constructs in rural environmental studies.

2.2.2.2 A Dynamic View to Research on Rural Built Environment

The transformation processes of rural built environments remain underexplored across both settlement-scale and building-level analyses. Prevailing scholarship predominantly examines spatial configurations as fixed outcomes, neglecting the formative dynamics that critically determine how and why these environments emerge. This methodological gap obstructs comprehension of rural spatial evolution.

People constitute the catalytic variable shaping built environments. As both creators and users of spatial systems, communities fundamentally govern all transformation processes. Through daily inhabitation patterns and incremental modifications, residents continuously imprint socio-cultural values onto material landscapes, generating distinct architectural variations.

Contemporary rural scholarship increasingly emphasizes dialectical frameworks that reconcile historical continuity with contemporary adaptation while simultaneously bridging static material analyses and developmental process tracing. To fully apprehend architecture's mutable essence—particularly in quotidian environments—this study advocates processual methodologies that capture dynamic interactions between spatial forms and temporal flows.

2.3 Research Data: Literature, Archive and Field Investigation

The data needed in the research mainly rely on the first-hand materials obtained in the field survey. During the field study, several approaches are used, including surveys on architectonic remains built in different periods, interviews with villagers and local officers, and the collection of local historical literature such as family genealogies and land deeds. Indeed, the field survey is not merely limited to the village. The surrounding areas of the village are also worthy of observation. They provide valuable background information to rethink the phenomenon and enhance the study of the village case reciprocally. These family archives provide precious information about the families who live in the village.

Other resources are official archives, such as local chronicles, local annals, population censuses, and economic statistics. This kind of data published by the government indicates the general development of the village and official policies that influence the local built environment.

The third data is from the survey of architectonic remains. It is a basic method to obtain architectural data directly. All the buildings are recorded and photographed with the following items like built time, material, shape, form, and ownership. It provides sufficient evidence to reconstruct the development process of the village and visualize the transformation of the built environment. The visualization of architectural data relies on the morphology of settlement pattern and typology of building fabric.

In this research, a cross-reference between rural social history and the built environment is conducted by comparing the architectural data from the survey to the literature data from various archives. It illustrates what (content), why (cause), and how (approach) the transformations happen in Xiaqiao.

2.4 Research Significance

The significance of this research lies in its methodological contribution, empirical findings, and practical applications for the rural built environment.

First, the study introduces an integrated multi-level analytical framework that encompasses time, space, and people. Unlike traditional approaches that focus primarily on shifting architectural styles or isolated spatial forms, this framework examines the built environment as a dynamic system. By analyzing the interaction between macro-level institutional mandates and micro-level adaptations across seven levels, the research provides a systematic method for understanding the complexity of rural spatial evolution.

Second, using Xiaqiao Village as a representative case, the study provides a comprehensive record of common rural structures. Many vernacular buildings in rural China remain in active use, and new constructions often continue to employ traditional techniques alongside modern materials. This research generates empirical knowledge on these underrepresented structures, documenting how indigenous traditions persist or transform during modernization. Such data is essential for understanding the broader socio-spatial changes occurring across rural China since the 17th century.

Third, the findings offer an analytical basis for interventions in rural planning and architectural conservation. The study provides specific references for guiding repair works and new construction, particularly within the context of Taishun. By clarifying the logic of spatial transformation and the governance hierarchies involved, the research assists policymakers and practitioners in balancing heritage preservation with sustainable development, offering a framework for evidence-based decision-making in rural revitalization.

Chapter 3
Research Methodology

This chapter delineates the methodological framework underpinning the investigation of rural built environments. A seven-level analytical structure is developed and operationalized through the case study of Xiaqiao Village, enabling multiscalar examination across varying observational parameters. This framework establishes the theoretical foundation for subsequent analytical chapters.

The discussion commences with an epistemological examination of the "level" concept in architectural research, situating the study within established multi-layered analytical paradigms. Foundational theories informing the methodology include urban morphological approaches pioneered by M.R.G. Conzen and Open Building Theory articulated by N.J. Habraken. These theoretical constructs facilitate spatiotemporal decomposition of built environments through three critical dimensions: stratigraphic analysis of historical layers, systemic disaggregation of spatial components, and agency-driven transformation processes.

Responding to China's rural modernization challenges, the chapter synthesizes these theoretical premises into a novel seven-level analytical framework. Empirical validation occurs through application to Xiaqiao Village, with particular emphasis on operationalization protocols for each analytical tier, context-specific evaluation metrics development, and inter-level relationship mapping methodologies.

The framework's implementation demonstrates enhanced capacity to decouple physical transformations from socio-economic drivers, quantify vernacular adaptation mechanisms, and establish heritage conservation priority indices.

3.1 Multi-level Theory in Built Environment Research

A level is a way to describe the spatial features of the built environment from the observation lens with different resolutions. It is used to define space with different scales, from the macro scale to the micro scale.

X. Lin and H. Wang, *A Multi-Level Analysis of China's Rural Built Environment*, Urban Sustainability, https://doi.org/10.1007/978-981-95-4945-0_3

The built environment is a complex system that integrates spatial forms at different scales. It encompasses various elements, including settlements, streets, plots, and buildings, among others. All types of built environments are characterized by a high degree of complexity, which is attributable to the continuous transformations they undergo. The built environment is not static; rather, it is a dynamic spatial cluster consisting of configurations with three dimensions: time, space, and people. The entity under consideration is formed by space, controlled by people, and exists over time. The three dimensions are interrelated and mutually transformable. It has been demonstrated that spatial configurations are susceptible to modification by disparate groups of individuals across varying temporal domains. The repercussions of temporal, spatial, and human forces manifest as visible physical remains in the built environment, thereby giving rise to the unceasing and inevitable transformations that characterize this environment.

There are different approaches to defining levels. Morphological research and Open Building theory provide the most important methods to layer the built environment.

3.1.1 Level Method in Morphology Research

The theory of urban morphology originated from the study of human geography in Germany. The origin of the theory can be traced back to the research of morphogenesis by the German geographer Schlüter in the late nineteenth century. He first proposed the concept of morphology of the "cultural landscape," pointing out that the first priority of research into human geography is to study the evolutionary process of the physical cultural landscape. This does not only emphasize the study of the characteristics and distribution of artificial entities, but also explores the origin and historical development based on natural, humanistic, and historical backgrounds. A thorough understanding of urban morphology is based on a comprehensive understanding of three elements, interrelated form (form), function (function), and the development process (development) [1].

The theory of urban morphology is predicated on the consideration of urban forms as well as the agents and processes that drive their transformation [2]. The objective of morphological study is to improve understanding of the built environment as a complex physical entity [3]. The aforementioned theory is comprised of three distinct aspects. The initial step in this process is the construction of a descriptive system for the spatial form, encompassing elements such as plan, street, plot, land use, building, and building fabric. This spatial hierarchy constitutes the foundational framework for delineating spatial domains across varying scales [4–6]. Secondly, it establishes a set of theories to explain the formation and transformation processes of urban space from different perspectives. These perspectives include the relation between form and function, the formation of landscape, the role of the decisive agent, and the economic influence [7–8]. Thirdly, it has been demonstrated to assist in the preservation of the aesthetic quality of historical urban spaces, while concurrently facilitating the adjustment of spatial transformation [9–10].

There are four principal approaches to the study of urban morphology [11–12]. These are: the historical-geographical approach, the process typological approach, space syntax, and spatial analysis. The historical-geographical approach originates from the German study of human geography and was further improved by Conzen into the fringe-belt, the morphological region and the burgage cycle. This approach was developed by the Urban Morphology Research Group (UMRG) represented by Jeremy Whitehand. He developed an explicit concern for agency in the fringe-belt process, undertaking studies that probed the interactions between landowners, developers, financiers and planners in the land use and development processes [13–15].

The process typological approach was developed by Saverio Muratori and Gianfranco Caniggia. The text addresses the principles of urban growth and transformation over an extended time period. The fundamental themes of territory interpretation and typological process are examined. The primary focus of this study is on the architectural group's growth within the context of urban spaces [16].

The space syntax approach is a more quantitative method to research the urban form. It was developed by Bill Hillier mainly to understand how architectural design influences the existing housing problem [17–19]. Space is the Machine is a book that provides a synthesis of the development of space syntax [20]. In 1998, the book Decoding Homes and Houses showed how domestic space provides a shared framework for everyday life.

The spatial analysis approach includes three models that are cellular automata, agent-based models, and fractals. These three models are complementary to each other. Michael Batty is the key researcher promoting this approach. In the book Cities and Complexity [21], he explains the complexity of city regions, using cells as a model to represent plots. This approach is more heterogeneous than the others mentioned above.

3.1.1.1 Level

The framework of built levels is the theoretical foundation of the urban morphology theory developed by Conzen. In the 1960s, Conzen, M. R. G., was widely regarded as having been instrumental in the development of the discipline of historical human geography in the UK, which had been initiated prior to the Second World War. The comprehensive theoretical system he developed was named urban morphology and is now widely known as the Conzenian School. The Conzenian School (also referred to as the British school) is considered to be one of the most prominent schools of urban morphology [22]. It is primarily associated with the historical-geographical approach [12].

In morphological research, the analysis is carried out based on different levels of the built environment [5]. Townscape is comprised of three elements at different scales: town plan, building fabric, and land utilization. These factors are relatively independent but closely linked.

The town plan is the most stable. The structural integrity of the edifice is found to be less stable in comparison to the urban layout. Land utilization is the most changeable. In comparison with other sources, the historical information contained in the town plan is the most abundant and thorough. Consequently, it is imperative to undertake a thorough and detailed study of the town pattern, with the objective of tracing its formation and development.

3.1.1.2 Process

Besides built levels, morphological research emphasizes the analysis of the transformation process. Rather than being studied as a constant state, urban spaces are studied as the result of "urbanisation as a process" [23]. This process is generated through a series of interacting factors, including society, economy, population, politics, culture, productivity, and the environment. The urban form is understood as an outcome of urbanization. In recent studies, complexity and uncertainty are emphasized in the definition of space. [21, 24].

The evolution of urban landscapes has been thoroughly conceptualized and refined. The evolution of material forms is inextricably linked to the development of society and the economy. The physical and spiritual differences that characterize different periods provide the necessary conditions for cultural presentation. The topography of the town has been shaped by the gradual accumulation of various cultural periods, resulting in a unique landscape that reflects the historical evolution of the region. In various geographical regions, the sequence and content are often influenced by the cultural period, and this phenomenon is also observed in urban areas. From a geographical perspective, those periods that leave unique material remnants in the landscape can be defined as morphological periods. The plan pattern and the building entity serve as the tangible carriers of morphological periods, and different morphological periods yield distinct material morphological characteristics.

The role played in the transformation process has led to the establishment of a morphological framework. This synthesis is the culmination of meticulous layering and diachronic research methodologies. The morphological framework refers to the elements that were formed early on and influenced and restricted subsequent development. Examples of these elements include the early geomorphic contour lines and the planar features of the initial construction. These morphological elements have exhibited stability and continuity in the development and changes of towns. In the formal factors, the town plan demonstrates the greatest stability. The town's developmental trajectory is profoundly influenced by its morphological framework, which comprises two substructural elements: the block and road system. Despite the town's significant transformations, vestiges of its former land and road boundaries often persist in the newly formed topography.

In conclusion, the urban morphological theory defines the composition of the morphological objects in a rigorous system, emphasizing the conceptual understanding of the morphological evolution process. The urban morphological theory provides a rigorous system for defining the composition of morphological objects, with an emphasis on the conceptual understanding of the morphological evolution process. Previous studies offer reliable approaches for research on forms at different levels. However, most of them are limited to formal analysis instead of a comprehensive interactive result of people and space. In terms of research interests, these works mainly focus on the spatial pattern of the outdoor environment, and the indoor spaces of building entities are seldom discussed. As there are great disparities between urban and rural contexts, it is useful to consider how the level theory of urban spaces can be applied to the research on rural settlements in China.

3.1.2 Level Method in Open Building Theory

The level method has been greatly improved in the Open Building Theory, which is used to elaborate on the relationship between the built environment and transformations in depth.

The concept of "Open Building" has been the subject of discussion for around thirty years, but related research has taken place for over one hundred years. In 1914, Le Corbusier designed the model of the Dom-Ino House, which proposed an open floor plan supported by reinforced concrete columns. This separates the structure into functional spaces and is seen as the primary prototype of the open building concept.

Systematic research into the open building theory commenced from John Habraken's "support-infill" concept [25]. He elaborates on this in his book *Supports: An Alternative to Mass Housing* that was published in 1961. From 1965 to 1975, he developed the support/infill method for the design and construction of adaptable housing while he held the position of director of SAR (Stichting Architecten Research) [26]. Habraken tends to define structures and other architectural components with non-structural functions separately. He distinguishes these two categories of building components by their ability and sensitivity to adapt to changes in the environment. This aims to solve the contradictions between the issues of efficiency and diversity, universality and individuality.

The open building theory is a methodological framework employed in housing design to direct the research on building entities. Subsequently, it evolves into a broader context, examining the transformation from environment to building [27–28]. The application of support/infill in the context of the environment involves the examination of the environment as shaped by acts of transformation, thereby revealing the elements that remain constant. This is the ideal setting for the exploration of the relationship between transformation and continuity in form.

The open building theory stratifies three levels at different scales, namely neighborhood as the tissue level, building as the support level, and inner space and furnishing as the infill level. These three levels follow different rules of transformation from the most stable situation to the most sensitive situation. Besides, the function of people, as groups or individuals who control specific space, is emphasized as well. These standards provide valuable insights to study the spatial transformations.

In the masterpiece written by Habraken, The Structure of Ordinary [29], the research is not merely limited to the architectural sense or urban form. He mentions the method to study rural settlement and the possibility of its application even in researching the Chinese rural environment. A lot of examples of traditional Chinese buildings are used to elaborate some of his ideas.

Configuration is a term applied in Habraken's work to describe a set of built elements in space. He further defines live configuration to emphasize the impact on the built environment under control [30]. Thus, live configuration is actually a specific grouping of built elements under an active unified control.

In the theory, three orders have been defined containing form as the physical order, place as the territorial order, and understanding as the cultural order. In different orders, a spatial hierarchy exists at different spatial levels. It dictates rules to control built elements such that configurations of lower levels are often separated and delimited by higher levels [29–30].

Moreover, people, as the controlling agent, are imported into the game of space control. The dimension of the built environment has been advanced into a complex of time, space, and people. It becomes a complex game, played by far more players on many levels over an extended period.

From the perspective of architecture, built environments at multiple scales are studied, with multiple parameters undergoing continuous changes. The study of transformations is of prime importance in understanding the built environment as a living, dynamic entity. Habraken defines a rather precise formulation on transformation [29–31].

In recent decades, the open building theory has been further developed by other researchers from different perspectives. Stephen Kendall focuses on the adaptability of residential buildings [32–33]. He researches open building theory in the context of new technology, new logistics, and new design methods. Jia Beisi explores the potentiality of adaptable housing especially in the context of the urbanization of China [34]. He also conducts research from a historical perspective to study the flexibility of traditional housing systems in China. Meanwhile, he is involved in practice with the BE architecture cooperation to develop the potential of mass housing [35].

The open building theory is predicated on a multi-level system. The built environment, comprising both the outdoor environment and building entities, is explored through a systematic breakdown into distinct levels. This breakdown encompasses the tissue level, the support level, and the infill level, with each level corresponding to a specific degree of adaptability and changeability within the spatial context. The text delves into the hierarchy of the control group on spaces on different levels. Synthesizing the results from the three dimensions of time, space, and people enables a more profound examination of the transformation of the built environment.

However, due to the limitations of context, extant research into open building theory in Western countries has focused on urban spaces, considering the city rather than the countryside. In China, the countryside presents considerable scope, and

China's urbanization process is surrounded by and interwoven with rural problems. The challenge of building in the rural environment and developing rural society has become one of the most serious issues in Chinese society and necessitates further research.

3.2 Construction of Seven-Level Research Framework

3.2.1 Seven-Level Research Framework

Scopes influence what we see and feel. They relate to the caliber of probes to observe the world and identify phenomena, and the degree of accuracy to describe and measure objects. By the multi-level approach, a synthesized result can be concluded that enhances our sense of the rural built environment thoroughly, on the topic of what occurred, is occurring, and is going to occur in rural China.

In this research, a seven-level analytical framework is established by grouping all the information into seven categories, according to different scales of observation. The seven levels are the first regional level, the second settlement level, the third sub-settlement level, the fourth neighborhood level, the fifth building group level, the sixth individual structure level (enveloping, structure, and space), and the seventh infill level (infill, furnishing, and equipment), following the sequence from macro to micro scope.

3.2.2 Define the Seven Levels in the Research of the Xiaqiao Village Case

The Xiaqiao Village is used as a case study in this research. The village is a temporal and spatial mixture. The architectural antiques with the longest history still standing in Xiaqiao are one and a half buildings built at the turn of the Ming and Qing dynasties, around the seventeenth century. It is the earliest image of Xiaqiao's built environment that can be traced, articulated, and shown through reliable visual evidence. Before that, due to the vulnerability of traditional wood or adobe construction to various natural and social elements, notably water, wind, pests, fire, wars, and other disasters, no architectonic remains built before the Ming dynasty exist on the ground now. Thus, instead of a specific building, the history of the earlier period, spanning more than 500 years, is transferred to communal memories, family beliefs, and local rituals. This history continues to affect the locals and their living spaces in an intangible way. In the past century after the imperial period, the dramatic changes in Chinese society as a whole have simultaneously affected vernacular rural landscapes. By transforming social impacts into built elements, temporal information from different times leaves its own imprints on spatial fragments.

Based on the complex composed of built environments and transformations that have occurred, both of which have a multi-level structure, a multi-level model of analysis needs to serve as a framework to research the interrelations among time, people, and space.

The diagram shows the seven levels in the case of the Xiaqiao Village (Fig. 3.1).

Transformations occurring at different levels have distinct features. The first regional level studies the general cycle of settlement development and population movement in a larger geographical field; second settlement level studies the sectorization and layout pattern of the village as a whole; third sub-settlement level studies the boundary and shape of sub-settlement; fourth neighborhood level studies the approach of the combination of several buildings as a group; fifth building cluster level studies the configuration of a building compound that was used as an independent house; sixth individual structure level studies the form, structure and indoor spaces of the building; seventh infill level studies the configuration of partitioning, furniture and equipment inside the building.

The first three levels belong to levels of settlement, while the last four levels belong to levels of architecture. A higher level (like the first level) implies the general trends of rural development that are more controlled by governmental powers. It is the result of collective actions. Meanwhile, a lower level (like the seventh level) indicates the daily life of ordinary people. It depends more on personal experiences. Furthermore, a hierarchy is implied in the seven levels that higher levels can influence the transformations on lower levels. However, the impact is transmitted in a one-way manner, meaning that lower levels have no influence on higher levels.

The competition and cooperation among different agents with governance powers at different levels, which lead to rural changes, are one significant concern in my research. It directs the discussion toward the hierarchy of different built levels that influence, constrain, and even dominate each other. It results in a mutual relationship of governance.

3.3 The Structure of the Two-Volume Set

3.3.1 Scope of Volume I

Volume I of this series focuses on the first three levels of the analytical framework: Level 1 (Regional), Level 2 (Settlement), and Level 3 (Sub-settlement). These levels collectively represent the macro-spatial scale of the rural built environment. The analysis at this scale investigates the physical form, spatial distribution, and developmental trajectory of the village as a collective entity.

Level 1: Regional Level examines the village within its broader geographical and administrative context. It analyzes how regional factors, such as topography,

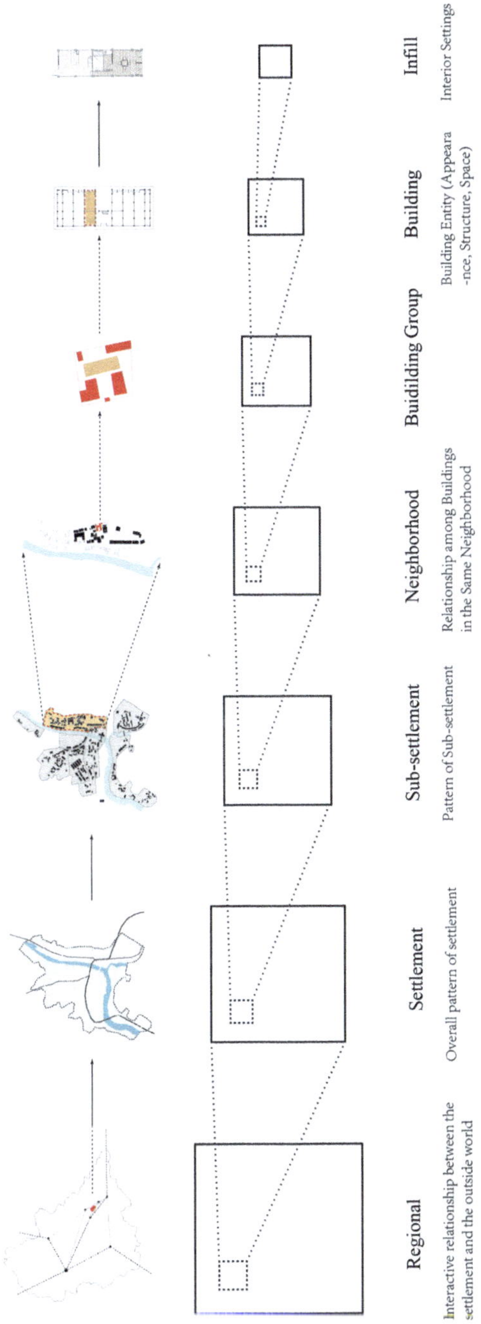

Fig. 3.1 Seven-level framework of the Xiaqiao Village

river systems, and government administrative zoning, determine the location and external connectivity of the settlement.
Level 2: Settlement Level investigates the village as a complete spatial unit. It focuses on the overall layout, the definition of boundaries, and the distribution of land use. This level analyzes how the settlement's morphology is shaped by collective forces, including lineage organizations and municipal planning.
Level 3: Sub-settlement Level looks at the distinct zones within the village. It examines how the settlement is subdivided into smaller residential or functional groups. The analysis focuses on the formation of these zones and their evolution from lineage-based territories to function-based districts.

By concentrating on these three levels, Volume I establishes the structural and institutional context of the village. It defines the physical boundaries and the governance systems that regulate the collective environment, providing the necessary framework for understanding the micro-level adaptations discussed in Volume II.

3.3.2 Scope of Volume II

Volume II focuses on the remaining four levels of the framework: Level 4 (Neighborhood), Level 5 (Building Cluster), Level 6 (Individual Structure), and Level 7 (Infill). These levels represent the micro-spatial scale, dealing with the specific living environments created and used by residents.

Level 4: Neighborhood Level examines the spatial relationships between adjacent buildings. It analyzes how individual houses are grouped and how semi-public spaces between them are organized.
Level 5: Building Cluster Level investigates the configuration of buildings within a specific lot. It explores housing typologies, such as courtyard houses and row houses, and how they relate to family structures.
Level 6: Individual Structure Level analyzes the physical entity of the building. It focuses on the architectural form, the load-bearing structure, and the spatial volume.
Level 7: Infill Level examines the interior configuration. It includes the arrangement of non-structural partitions, furniture, and equipment used in daily life.

This volume investigates how individual families and villagers operate within the macro-spatial framework established in Volume I. It analyzes the specific construction behaviors and adaptive strategies employed by residents to modify their immediate living spaces.

3.3.3 The Nested Relationship Between the Volumes

The seven levels of the analytical framework form a nested system. Each higher level provides the physical and institutional context for the level below it. Consequently, the macro-spatial levels analyzed in Volume I (Region, Settlement, and Sub-settlement) establish the structural framework and constraints for the micro-spatial levels analyzed in Volume II (Neighborhood, Building Cluster, Individual Structure, and Infill).

The relationship between the two volumes is defined by the hierarchy of control. The macro-spatial environment in Volume I is primarily determined by collective agents, such as the government and lineage organizations. These agents define the boundaries of the settlement, the layout of infrastructure, and the allocation of land. These decisions create a set of physical and regulatory limits.

The micro-spatial environment in Volume II operates within these limits. Individual families and residents adapt their buildings and interiors to fit the available land and comply with, or sometimes resist, external regulations. Therefore, the spatial patterns observed in Volume II are not independent; they are responses to the conditions established in Volume I. Understanding the full transformation of the rural built environment requires examining both the macro-level framework and the micro-level adaptations as an interconnected process.

Refereneces

1. Whitehand J. W. R. (1981). Background to the urban morphogenetic tradition. In J. W. R. Whitehand (Ed.), The Urban Landscape: Historical Development and Management. Academic Press. pp. 2.
2. Oliveira V. (2017). Urban morphology: An introduction to the study of the physical form of cities. Raumforschung und Raumordnung|Spatial Research and Planning, 75(3).
3. Kropf K. (2018). The handbook of urban morphology. Wiley.
4. Carvalho Rui and Alan Penn. (2004). Scaling and universality in the micro-structure of urban space. *Physica A: Statistical Mechanics and its Applications* 332:539–547.
5. Conzen M. R. G. (2004). Morphogenesis and structure of the historic townscape in Britain. In: Conzen MRG, Conzen MP (eds) *Thinking about Urban Form: Papers on Urban Morphology 1932–1998*. Peter Lang, Oxford.
6. Cooper J, Oskrochi R. (2008). Fractal Analysis of Street Vistas: A Potential Tool For Assessing Levels of Visual Variety in Everyday Street Scenes. *Environment and Planning B: Planning and Design* 35(2):349–363.
7. Caniggia G., Maffei G.L. (2001). Architectural composition and building typology: interpreting basic building. Florence: Alinea Editrice.
8. Brown C., Witschey W. (2001). The Fractal Geometry of Ancient Maya Settlement. *Journal of Archaeological Science* 30:1619–1632.
9. Conzen, M. R. G. (2008) Retrieving the Pre-industrial Built Environment of Europe: The Historic Town Atlas Programme and Comparative Morphological Study. *Urban Morphology* 12:143–156.
10. Conzen, M. P. (2009). How cities internalize their former urban fringes: A cross-cultural comparison. Urban Morphology. 13 (1): 29–54.

11. Moudon, A. V. (1997). Urban morphology as an emerging interdisciplinary field. Urban Morphology, 1(1): 3–10.
12. Kropf, K. (2009). Aspects of urban form. Urban Morphology, 13(2): 105–120.
13. Whitehand JWR, Morton NJ. (2003) Fringe belts and the recycling of urban land: an academic concept and planning practice. *Environment and Planning B: Planning and Design* 30:819–839.
14. Whitehand JWR, Morton NJ. (2004). Urban morphology and planning: the case of fringe belts. *Cities* 21:275–289.
15. Whitehand JWR, Morton NJ. (2006). The fringe-belt phenomenon and socioeconomic change. *Urban Studies* 43:2047–2066.
16. Cataldi G, Maffei GL, Vaccaro P. (2002). Saverio Muratori and the Italian school of planning typology. *Urban Morphology* 6:3–14.
17. Hillier B. (1973). In defense of space. *RIBA Journal (Royal Institute of British Architects Journal)* 539–544.
18. Hillier B., Leaman A., Stansall P., Bedford M. (1976). Space syntax. *Environment and Planning B: Planning and Design* 3:147–185.
19. Hillier B. (1996). Cities as movement economies. *Urban Design International* 1:41–60.
20. Hillier B. (1996). Space Is the Machine. Cambridge: Cambridge University Press.
21. Batty Michael. (2005). Cities and Complexity: Understanding Cities with Cellular Automata, Agent-based Models, and Fractals. Cambridge, MA: The MIT Press.
22. Huo Y霍耀中, Gu K谷凯 (2005) Shizhen guihua fenxi gainian, fangfa yu shijian 市镇规划分析:概念、方法与实践 (Concepts, methods, and practices of town planning analysis). *Urban Development Studies* (2):27–32.
23. Knox, Paul. (1994). *Urbanization: An Introduction to Urban Geography*. Prentice-Hall, Englewood Cliffs, N.J.
24. Batty Michael. (2008). Fifty years of urban modeling: macro-statics to micro-dynamics. In: Albeverio S, Andrey D, Giordano P, Vancheri A (eds) *The Dynamics of Complex Urban Systems*. Physica-Verlag HD, Heidelberg. https://doi.org/10.1007/978-3-7908-1937-3_1.
25. Habraken, N.J. (1961) *De Dragers en de Mensen: Het Einde van de Massawoningbouw*. Scheltema & Holkema, Amsterdam.
26. Bosma, Koos, Dorine Van Hoogstraten, and Martijn Vos. (2000). Housing for the Millions: John Habraken and the SAR (1960–2000). Rotterdam: Nai Uitgevers.
27. Habraken, N. J. (1972). Supports: An Alternative to Mass Housing. London: Architectural Press.
28. Habraken, N. J. (1983). Transformations of the Site, Cambridge: Awater Press.
29. Habraken, N. J. and Jonathan. Teicher. (1998). The Structure of the Ordinary: Form and Control in the Built Environment. Cambridge, Mass.: MIT Press.
30. Habraken, N. J. (1987). Control hierarchies in complex artifacts. In Proceedings of the 1987 Conference on Planning and Design in Architecture. pp. 75–84.
31. Habraken, N. J. (1985) The Appearance of the Form, Cambridge: Awater Press.
32. Kendall, Stephen. (2000) Residential open building. Edited by Jonathan Teicher and Studies and Documentation International Council for Building Research. New York: Ltd. Spon Press.
33. Kendall, Stephen H., and Jonathan Teicher. (2010). Residential open building. Abingdon, UK: Routledge.
34. Jia, Beisi. (1995). Adaptable housing or adaptable people? *Architecture et Comportement/ Architecture and Behaviour* 11:139–162.
35. Jia, Beisi. (2001). Flexible housing, compact city and environmental preservation: a critical look at Hong Kong's experience. *Open House International* 26(1):26–33.

Chapter 4
The Up-Down Kingdom of Formal Governance: Mapping the Settlement Levels

This chapter examines the transformation of the built environment in Xiaqiao Village through the lens of settlement morphology, conceptualized as a complex interplay of buildings, spaces, and landscape. The presentation of the different settlement configurations follows an order of increasing resolution of observed objects. It starts with a description and explanation of the regional relationship between Xiaqiao and other related settlements. It then increases the resolution and moves to the village as a whole settlement, independently, and then to several sub-settlements that constitute Xiaqiao, analyzing the intrinsic properties of spatial patterns.

The morphology method is used in this chapter as the main research approach to analyze the form, patterns of spaces, and landscapes. These morphological diagrams explore the process of transformation throughout the four different socio-economic periods, based on various influencing factors and evaluation criteria. In the last part, the nexus of control is further explained. This nexus defines the central operational relationship between humans and the dynamic patterns of change that echo throughout a built environment, in order to reveal the complexity, flexibility, and dynamic pluralism of the rural built environment in China.

4.1 Level 1: Regional Level

The first regional level is the highest level to describe the built environment with the largest observation scale. At this level, the whole village is abstracted to a point to explain the interaction between the selected housing settlement and the outside world. All the transformations of the built environment that happen at the regional level follow general trends that provide a basis for what occurs at the remaining seven levels. These changes can be seen as norms formulated at this level, which trigger a series of construction activities and then regulate subsequent construction activities.

X. Lin and H. Wang, *A Multi-Level Analysis of China's Rural Built Environment*, Urban Sustainability, https://doi.org/10.1007/978-981-95-4945-0_4

The spatial configuration at the regional level can be translated into spatial representations through two approaches: the geographical pattern and the social pattern. The space represented by the geographical pattern describes the physical environment predetermined by nature. Meanwhile, the space represented by the social pattern describes the political, economic, and cultural conditions. These conditions are influenced by the state government and shaped by historical trends.

4.1.1 Mapping the Geographical Features: Preconditions of Settlement Development

The geographical environment exerts a profound influence on regional culture and local lifestyles. While the deterministic view that geography solely shapes culture, known as geographical determinism, is an overstatement, it remains undeniable that geographical features fundamentally condition the phenomena and characteristics of human activity and settlement patterns. The general development trend cannot escape the constraints that nature sets in advance.

The natural context is the first condition that affects how a settlement is established. It contains the land relief, quality, and suitability of soils and subsoils, climate, solar and wind exposure, and the type of natural landscape. All these aspects provide expression and manifest in forms [1]. Geographical factors, which are physical and biological factors tied to geographic location, include climate, distributions of wild plant and animal species, soils, and topography. These factors dictate options for artificial products such as clothing, shelter, and food.

The landscape pattern, constituted by continuous mountains and relatively independent water systems, is the main structural framework that shapes the spatial pattern from the perspective of geography. It is the crucial factor that causes the differentiation between southern Zhejiang and the rest of the province. To analyze the geographical pattern logically, three levels are presented, starting from a broader geographical unit, the southeast coastal mountains (SCM, 东南沿海丘陵), gradually zooming in on the topography of Taishun County, and finally on the specific landscape where the village is located (Fig. 4.1).

4.1.1.1 Southeast Coastal Mountainous Area in China

SCM is the major geographical unit in southern China. SCM, located in southeastern China, constitutes a distinct natural geographical unit. It is characterized by its unique combination of subtropical mountain and coastal landscapes and includes all parts of Fujian province and some parts of Zhejiang and Guangdong provinces (Fig. 4.2). The terrain here slopes down from northwest to southeast, with the mountainous area accounting for about 85% of the total area of the region [2].

Fig. 4.1 Left: The Field of SCM in China; Right: The Map of Southeast Coastal Mountains Drawn by Skinner [3]

Fig. 4.2 Diagram of the Marginality of Taishun

Continuous mountains separated by deep valleys and connected by intersecting rivers are the typical topography found in SCM. On one hand, mountains make SCM an independent geographical unit with a culture and living customs different from those in other neighboring areas, such as the plains along the Yangzi River. On the other hand, the geographical similarities also foster similar characteristics of culture and social lifestyles. The cultural connectivity influences social development, resulting in interaction among different parts of SCM.

At first, the social development in SCM lagged behind that in neighboring plain areas due to a scarcity of farmland and inconvenient transportation. During the agricultural period, farmland was the major living resource that dominated economic development. But in the mountainous area, both the quality and quantity of arable lands were unsatisfactory. With the increasing population, the contradiction between humans and available farmland was aggravated. Moreover, during premodern times, the transportation system in the mountains could not be well developed, and the major means of transportation was by foot. It largely constrained the circulation of living resources as well as the communication of people. Thus, the two features produced by mountainous topography, insufficient arable lands and inconvenient transportation, resulted in the lagging development of society in SCM.

Secondly, it is hard to form a powerful governance unity in whatever corner of SCM, because numerous mountains and valleys cut the entire area into dozens of mutually independent geographical units. In contrast to the plain area, where the whole field can be connected efficiently in all directions, the topographical fragmentation hinders the establishment of an effective and highly unified official governance.

Furthermore, the two features produced by the unique geography, namely the difficulties of livelihood and the hardship of governance, contribute to an unstable social environment where frequent disturbances and continuous migration occur. Due to the poor economy and continuous disturbances in SCM, migration is a typical phenomenon. Residents have to move around looking for living resources. People prefer to move to the areas that are comparatively undeveloped with lower population pressure. Most of them just move to the neighboring areas due to the ease and cost of migration. Usually, Fujian is the place where people move out, due to the greatest pressure of living and the most serious contradiction between land and people. They tend to move to neighboring areas in southern Zhejiang, which is where people move in, due to the rather lower population density. Migration is so common that in Taishun, 74.2% of the people are immigrants who arrived after the Ming dynasty [4].

The migration culture also becomes a representative cultural feature in SCM. Instead of one-time behavior, migration behaviors happen as a series of chain reactions constituted by a group of successive events, and before they finally settle down, the migrants might continue moving. Culture and folk living customs are spread along the moving route and finally arrive at another place hundreds of miles away through the traveling of migrants. Therefore, migration, equipped with the unique culture, first fires up in one area and then spreads to the whole SCM area.

In general, due to the geographical obstruction and cultural similarities shared in SCM, although it is rather isolated from the outside, the inner parts of SCM influence each other highly and are sensitive to variations.

4.1.1.2 Geographical Environment of the Taishun County

Taishun is the county to which Xiaqiao Village belongs. Now we will discuss its geographical environment. Taishun sits at the junction of Zhejiang and Fujian provinces, which is the eastern boundary of SCM. It receives different social influences from eastern Fujian and southern Zhejiang. The marginality is also reflected in its location as the junction between inner mountains and coastal plains. Taishun is on the first step of the hilly area, and it takes only 1.5 hours by car to reach the coastal Cangnan County (Fig. 4.2). It is the transfer area connecting mountains and sea.

The topography of the entire Taishun is complex, full of steep hills, deep valleys, rivers, and basins. The terrain here is described accurately by the locals as "nine tenths of the land is mountains, half of one tenth is rivers, and the other half is farmland." The illustrations in all historical archives about Taishun can prove this [5–9]. The terrain of Taishun is greatly fragmented into dozens of small geographical units that are mutually independent.

The river system regroups the spatial fragments where rural housing settlements are located. Four rivers divide the whole Taishun County into four drainage areas: the Feiyunjiang River, the Jiaoxi River, the Shachenggang River, and the Aojiang River. The former two rivers go through most of the areas in Taishun, that Feiyunjiang in the northern part and the Jiaoxi in the southwestern part.

The river system plays a critical role in defining both the geographical and cultural circles. Inside mountains, the routes along rivers are more effective and efficient for daily communication, although some rivers are not navigable due to hydrological conditions. Thus, the connectivity of the river system shapes the formation of cultural communities.

The dialect circle in Taishun is one of the major cultural circles that have a great impact on people's daily lives. The dialects used by villagers in different parts have huge distinctions. There are mainly two dialect systems: the Wu dialect, used in Zhejiang Province, and the Min dialect, used in Fujian Province. Moreover, the Wu dialect can be classified into the Wenzhou dialect and the Liqu dialect; the Min dialect can be classified into the South Min dialect, the East Min dialect, and the Manjiang dialect [4].

The formation of dialect regions is related to the distribution of river systems. For example, in most of the areas along the Feiyunjiang River, people use the Wenzhou dialect due to the navigable connection along the river to Wenzhou City. However, in most of the areas along the Jiaoxi River, people use the Manjiang dialect, which belongs to the Min dialect, due to the connectivity with eastern Fujian Province by the river.

The complex mountainous topography obstructs transportation. It isolates the inner world from the outside. The complex terrain greatly increases the time and monetary costs and reduces the efficiency of transporting goods, transporting people, and circulating information. Besides, the available farmlands in the mountains are limited in both quantity and quality, making this area one of the poorest regions in Zhejiang Province for a long time [4]. Meanwhile, the development of Taishun was delayed compared with the neighboring areas. This left an area with a relatively small population, which then became a hot destination for migration during the Ming and Qing dynasties. Due to geographical proximity, most immigrants were from eastern Fujian or southern Zhejiang.

In general, the topography of Taishun is complex, full of mountains. The major geographical features are determined by the four river systems, which also influence the formation of regional culture.

4.1.1.3 Landscape of the Xiaqiao Village

Xiaqiao is located in the southeastern part, along the Jiaoxi River. The residents there use the Manjiang dialect, which is similar to the East Fujian dialect. It is close to the eastern part of Fujian geographically. The Xiaqiao Village is one of the earliest settlements that have been recorded in literature in Taishun.

The geographical location of Xiaqiao is favorable. Firstly, it is located in a small basin belonging to the central area of Sixi town. It is the only route for transportation between the five townships and the rest of Taishun, or from northern Taishun to the southeast corner or to its neighboring county (Fig. 4.3). The basin connects five adjacent townships and hundreds of villages, taking up an area of one third of the whole county. Due to its function as a regional traffic hub, it has become the center of the entire southeastern Taishun as well. The center of Sixi combined with the center of Yayang and Shiyang, as a whole, constitute the network of southern Taishun [10]. The formation and continuous development of the community all benefited from the advantages of its location. But its potential for development has also been limited in the southeast part for centuries due to the insular geographical environment and the undeveloped transportation system.

Regarding the Xiaqiao case, the surrounding landscape is also a predominant geographical feature here. The landscape surrounding Xiaqiao is special. Four rivers meander through the village and converge together. It forms one of the most fortunate

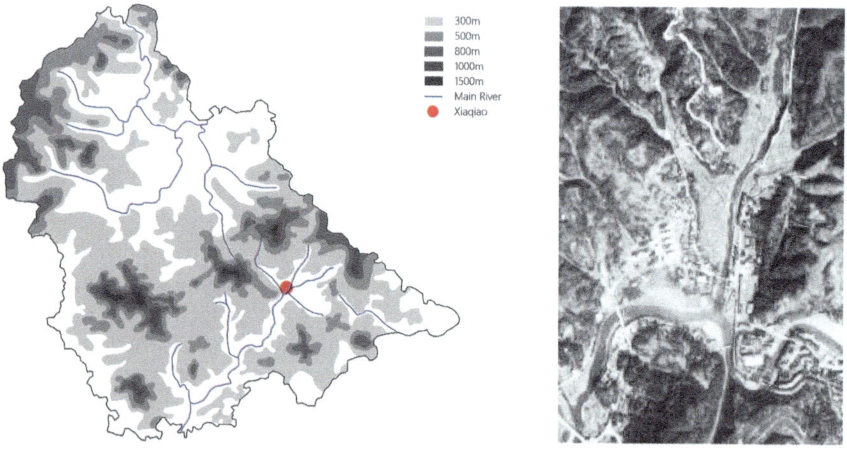

Fig. 4.3 Left: Location of Xiaqiao; Rirht: Landscape around Xiaqiao. (Redrawn by the author based on Google Map, 2017)

Fig. 4.4 Landscape of Xiaqiao

geomantic landscapes "Sishui Huilan" (Four rivers flow over, 泗水洄澜). Besides, a high mountain with three peaks named San Jiangjun Mountain (three generals, 三将军) sits on the south (Fig. 4.4).

The landscape constituted by water and mountains is seen as an origin of fortune that protects the village. It brings both prosperity and the risk of disaster to the village. The four rivers meandering here, surrounded by high mountains, are among the most favorable patterns in Fengshui (geomancy, 风水). The perfect pattern of the landscape attracts residents living here. But sometimes the large flow of water in the area of rivers' convergence leads to frequent flood disasters that happen frequently, resulting in the ruins of the village.

4.1.2 Mapping the Social Features: Background of Settlement Development

The general trend of social development, which delimits the potential range of the transformation of the rural built environment, cannot be escaped. Compared with the geographical features that are always treated as constants, the social features always evolve with the development of the whole society. Table 4.1 simply introduces Xiaqiao's history, which basically portrays the process of the village's development (Appendix I).

Besides, two crucial social features, namely the administrative network and the daily communication network, are selected to describe the major social changes that have the most significant influence on rural settlements. By mapping these two social features, an inter-nested hierarchy can be established to analyze the spatial patterning of human interaction.

Table 4.1 The General History of Xiaqiao [4, 10–11]

Period	Xiaqiao's development	Regional history background
The Five Dynasty	First be settled: The resident who has firstly settled here is sir. Lin Jian who seek for a quiet and peaceful life.	The whole China is in turmoil during the Five Dynasties. The phenomenon of migration is common during that period.
The South Song Dynasty	Booming: The descendants of the Lin lineage are successful in the official exams, and became officials.	The entire Wenzhou city became prosperous during the South Song dynasty.
The Yuan Dynasty	Depression: The Lin lineage has been declined.	During the yuan dynasty, the intellectuals from the South China were suppressed on political circles.
The Ming Dynasty	Start to revival, but still in a lower level: Wars leaded by South Ming dynasty; the immigration of the Tang and Chen lineages	The formal establishment of the Taishun county in the middle of the Ming dynasty. But in the turn of Ming and Qing dynasty, the disturbances were continuing. The village has been destroyed by the turmoil. Meanwhile, the unstable social environment is not good for wealth accumulation.
The Qing dynasty	Continuously develop	The whole society is in stable during the Qing dynasty. The techniques of handicraft have been advanced. New species of crop has been imported here.
The republican period	Stagnation	Continuous wars made the whole society undeveloped. But the whole Taishun area belongs to the old revolutionary base area that avoids the big scale of damage.
The planned-economy period	Communization	The whole society has been communized. The economy is undeveloped.
The market-economy period	Difficult	Comparing to the coastal plains of Zhejiang province, the development of the mountainous Taishun is lagging behind.
	Prosperous	The economic sector has been transitioning. The infrastructure has been updated. Along with the overall development of the mountainous area, the village has been developed further.

4.1.2.1 Spatial Pattern of Administrative Hierarchy

Xiaqiao is a village that is at the basic administrative level in China. Now it is governed directly by Sixi Town, which is one of the nineteen towns in Taishun County. However, compared to the villages located in peripheral regions, Xiaqiao undoubtedly occupies a central location. A larger-scale housing settlement has developed here. Xiaqiao sits inside a small basin and is an inescapable traffic hub connecting all five adjacent towns on higher mountains. But compared to its neighbor, Baifenqiang Village, which is exactly at the junction of several traffic routes, Xiaqiao is located more toward the northwest side of the basin. Thus, in the controversy over the town center location, Xiaqiao lost to Baifenqiang. The latter became the town center where the Sixi Town government is located. But from a geographical perspective, the town government office is located in the middle of the two villages near Xiaqiao, due to the consideration of Xiaqiao's historical and functional significance in Sixi Town. Thus, Xiaqiao can be seen as a sub-central village in Sixi Town with a larger scale than common ones.

4.1.2.2 Spatial Pattern of Xiaqiao's Accessibility

The accessibility of the village is decided by the development of the regional transportation system. During the Qing dynasty, there were four official roads[1] in Taishun, but Xiaqiao was not directly located on any of them. Transportation in the mountainous area mainly relies on foot. It takes an adult man a whole day to walk from Xiaqiao to the county's center, Luoyang.[2]

However, in the southeastern part of Taishun, Xiaqiao still is the traffic hub for the surrounding five mountainous towns, including Baoyang, Dongxi, Fengyang, Liufeng, Hengkeng. From the statistics of spouses' hometowns in the marriage relationships of the Tang family in the Qing dynasty, the daily communication circle always equals the area. This area is connected by the traffic hub Xiaqiao and covers Sixi Town and several surrounding villages (Fig. 4.5).

In 1958, a provincial road called 58 Road was built, but it did not connect to Xiaqiao directly [12]. Residents living in Xiaqiao needed to transfer in Liantou village, 15.4 km away. Then a town-level road, Xiaolian Road, that goes across the village, was built. It strengthened the role as a sub-regional traffic hub of Xiaqiao. The new upgrade of the whole regional transportation system did not appear until 2007. The new 58 provincial road was built, passing to the west of Xiaqiao directly. After that, the time for travel from the village to Luoyang or to Lingxi was greatly shortened to only 1.5 hours. Modern transportation methods bring great convenience

[1] Official Road (Chinese: 官马大道) It means the road built in premodern period in China by government for
transportation by horse riding

[2] Luoyang (Chinese:罗阳) is the center town of the Taishun county; Lingxi (Chinese:灵溪) is the center town of
the Cangnan County just next to Taishun

Fig. 4.5 Field of Daily Activity during the Qing Dynasty (see Appendix II)

to local residents. Today, the sphere of interpersonal communication has been greatly enlarged. Traveling by modern roads, the residents of Xiaqiao are able to access the high-speed railway station in Cangnan County and the airport in Wenzhou City conveniently. These efficient travel approaches take the local villagers to every corner across the country, even around the world.

4.1.3 Summary of the First Level

At the regional level, the focus is on research into the interaction between Xiaqiao and the external society, using the broadest observational scale. At this level, the entire village is abstractly represented as a point in order to explain the interaction between the selected housing settlement and the outside world. The various transformations of the built environment that occur at the regional level are influenced by overarching trends that serve as a foundation for the phenomena observed at other levels. These alterations can be seen as norms established at this level. As a result, they trigger a series of construction activities, which are then regulated.

At the regional level, the spatial configuration can be translated into spatial language through two approaches: geographical and social patterns. The space defined by the geographical pattern describes the physical environment presupposed by nature, whereas that defined by the social pattern describes the political, economic, and cultural conditions shaped by state policies and historical developments. The geographical environment establishes the overall constraints that influence the change of the artificial environment. Concurrently, the evolution of rural settlements is in line with the prevailing trends observed in the broader societal context. In contemporary times, the spatial distance between the village and the outside world has been significantly reduced.

4.2 Level 2: Settlement Level

The second level is the settlement level to describe the spatial pattern of the entire Xiaqiao settlement. It tries to explain the transformation of the rural built environment by treating the settlement as a complete object of observation, instead of focusing on partial internal interactions.

At this level, fourteen criteria are selected to measure the transformation of the settlement pattern, including shape, boundary, street network, occupancy situation, landscape, density, function, height, structure, material, construction period, architectural form, clan distribution, and ownership. By conducting a layered analysis of individual evaluation elements, the general result of settlement transformation can be synthesized.

In order to discuss the complexity of the formation of specific settlement patterns, a comparative study is conducted. It selects three typical rural settlement patterns in the SCM area that are 'Combed shape' village case in the Guangdong province, the 'ring shape' village case in the Fujian province, and the dispersed village case in the south of Zhejiang province, to explore how the different settings of nature, society and culture affect the formation of settlement patterns (Table 4.2).

4.2.1 General Configuration of Settlement

To analyze the general configuration of settlement, two criteria, the shape of settlement growth and the boundary of settlement, are used, in which the former describes the transformation of the process by which the settlement grows, and the latter describes the transformation of the approach by which the concept of 'ours' and 'others' settlement is defined on the settlement level.

Table 4.2 Analyzing element of settlement pattern

Analyzing objects	Criteria	Evaluating factors
General configuration (environment)	Shape	Centralized / de-centralized; Linear / conglomerated
	Boundary	Exist / disappear; Clear defined / gradually transitioned/mixed
Settlement's structure (environment)	Street network	Alley / Main street inside village / road going outside village
	Occupied situation	Built-up area / open space
	Landscape	Big tree / water / old well / mountain surrounding / agriculture landscape
Maturity of community (human beings)	Density	High/medial/low
	Function	Residential; shop; educational; Temple; bridge; storage; service; restaurant; hostel and hotel; mixed (residential and commercial);
Development of architectural technology (buildings)	Height	One floor / two floors / three floors / four floors / five floors / six floors
	Structure	Traditional wood structure; mixed structure I (wood frame structure plus stone/brick/rammed earth load Bearing Wall); mixed brick structure; brick and concrete structure; mixed structure II (brick walls plus concrete frame structure)
	Material	
	Built period	The End of Ming Dynasty; The Early of Qing Dynasty; The Late of Qing Dynasty; The Republican China Period (1912–1948); The Planned-economy Period (1949–1979); The Market-economy Period (1980–2016)
	Architectural form	
Property of Clan Village (human beings)	Clan distribution	The Houchi Lin lineage; the Huayuan Lin linage; the tang lineage; the Chen lineage; the Qiaotou Lin lineage; families from adjacent areas; families from other areas in Taishun; families form other area outside Taishun
	Ownership	Private / public / semi-public (shared by one lineage)

4.2.1.1 Shape

For the settlement shape, the configuration is formed by connecting the neighboring buildings with a distance of less than 30 m [13] (Fig. 4.6). The merging process of different parts inside the settlement is clearly illustrated. In the four periods,

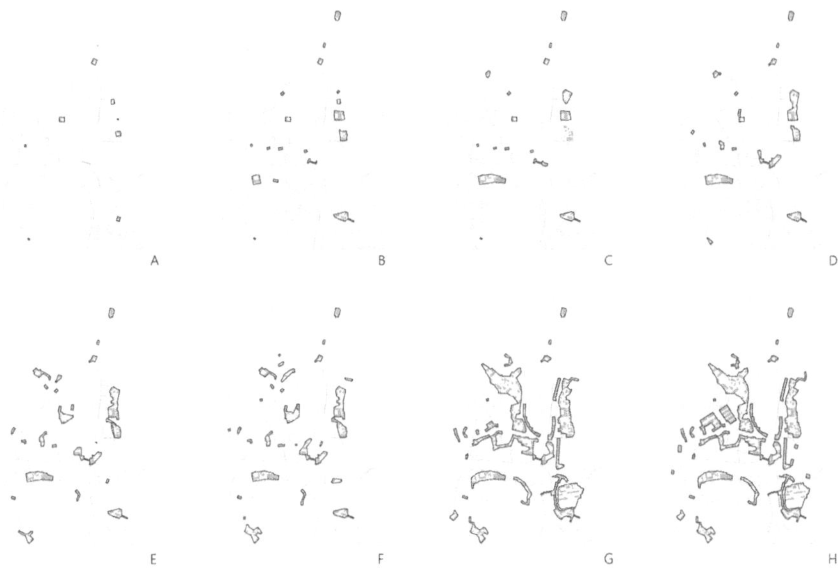

Fig. 4.6 Shape Transformation of Settlement (**a**) 1644–1820; (**b**) 1820–1911; (**c**) 1912–1929; (**d**) 1930–1949; (**e**) 1949–1969; (**f**) 1970–1979; (**g**) 1980–2005; (**h**) 2005–2016

the shape of settlement configuration can be generalized as disperse settled, group growth, cluster gathered, and linear connected.

In the Qing dynasty, the buildings were dispersed in the village, and residents settled in single-courtyard houses. Due to the rather stable population density, every two buildings kept a certain distance, surrounded by larger fields of farmland. Then, in the Republican period, more buildings were built around the original ones. As a result, small families moved out and built new houses due to population growth and the start of the disintegration of the big-family system. They always built new houses around the old ones due to homestead ownership. Thus, in the second period, the settlement grew in a grouped pattern. New territories were explored to satisfy more families. After the establishment of the PRC, the cluster effect was strengthened so that small housing groups evolved into bigger housing clusters. Some neighboring groups merged into a bigger cluster. But the population increase remained at a stable rate, and there are still many open spaces remaining inside the settlement. In the last period, along with the rapid economic increase, the settlement was totally merged into an integrated whole. But instead of growing by cluster, different parts of the settlement are connected by linear-shaped housing patterns along the new road, or resembling row-shaped modern housing units that are broadly used in urban spaces.

4.2.1.2 Boundary

The boundary of the village is the element that defines what is inside from what is outside. It is formed by natural elements that spontaneously parallel how the villagers recognize the surrounding environment. Along with the artificialization of the rural settlement, the boundary is redefined more by the built-up area. But with the continuous expansion of the built-up area, some parts of the boundary that differentiate between the two neighboring villages have disappeared (Fig. 4.7).

Natural elements always initially form the boundary of the settlement. The surrounding mountains and the circuitous rivers form the natural boundary of Xiaqiao, which is located in a basin. In the premodern period, Xiaqiao could only be recognized by conceptual features of the landscape. The locals usually say that the lands they can see in the village all belong to their village. Instead of an integral residential zone, Xiaqiao is actually made up of several independent small settlements. Each of them is obviously enclosed by geographical elements such as mountain ridges, rivers, and farmlands. The adaptability to the natural environment creates primary boundaries.

But after the expansion of the built-up area in Xiaqiao, the artificial boundary has gradually been created since the Republican period. The artificial boundary can be seen more clearly along with the growth of housing groups. It defines the field where human daily activities happen. This trend reflects the increasing separation between the built environment and the natural environment, highlighting the growing human-dominated transformation of the landscape.

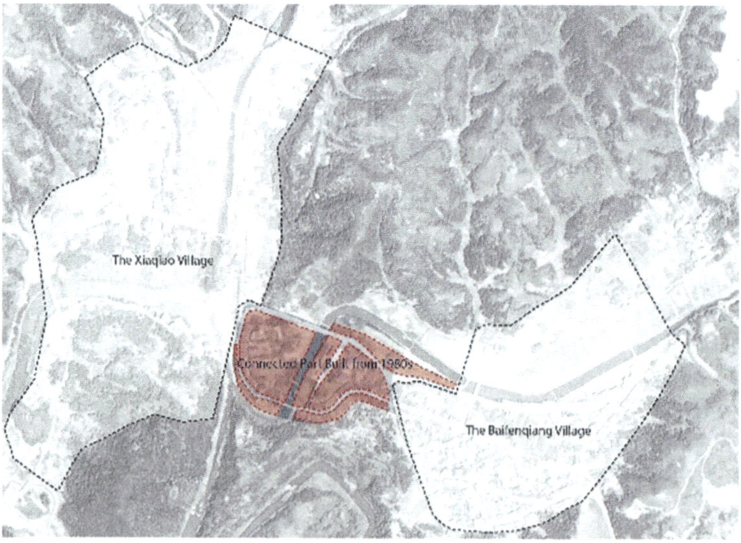

Fig. 4.7 Connection of the Adjacent Two Villages. (Redraw based on Google Map)

However, in the recent decade, the boundary has blurred again, and on the boundary between the two neighboring villages, it is hard to differentiate between ours and theirs. Take Xiaqiao and its neighbor Baifenqiang as an example. The Dongxi River is the original boundary between them. A highland named Tangqiangang naturally separates the two villages on the eastern bank of the river. But in the 1980s, the highland was split in the middle in order to build the Xinyang Road and the Sishui Road. The roads shorten the circulation time and efficiently connect the two villages. Both villages grow along the traffic lines to maximize the advantages of the roadside. They form a transitional residential area made up of mixed residents. Finally, the two villages merge together in the space around the Dongxi River. Thus, connectivity, replacing the previous geographical separation, blurs the spatial boundary again.

4.2.2 *Structure of Settlement*

To analyze the structure of the settlement, three criteria are used, namely street network, spatial occupation, and landscape. Street network is the framework of the settlement that regulates the flow of daily activities. Spatial occupation is the main body of the settlement that describes how buildings occupy the land in the village. Landscape is an additional value of the settlement that reflects how the village interacts with nature.

4.2.2.1 Street Network

The street is the route of daily circulation in the village. But it is not an original concept in a mountainous village with lower built density like Xiaqiao. There are only paths connecting houses and farmlands. The formal street network in the village is the result of governmental planning rather than a spontaneous development process. It appears along with the continuous construction inside the settlement after the first road, the Xiaolian line, was built in the 1970s (Fig. 4.8).

In the traditional path system, paths are formed spontaneously in response to the needs of daily circulation without prior government planning. In the Qing dynasty, the official roads in Taishun did not pass over Xiaqiao. There were only paths connecting houses and farmlands in the village. Compared to the big houses, bridges, ancestral halls, and temples, paths were not an evident element in the settlement and were not depicted on the village map in the family archives. With the construction of houses, pavement was created spontaneously toward the new house after it was built. It is clearly evident that the settlement's growth overlapped with the growth of the path network before the introduction of the modern street network.

Instead of forming the surface on the side of the path, the buildings tend to be treated as the destination of the path. It is another difference between the traditional path system and the modern street system. The surface along the path is always

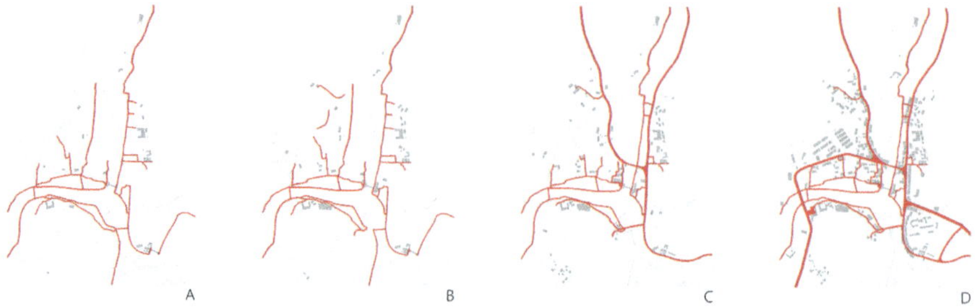

Fig. 4.8 Transformation of Street Network (**a**) 1644–1911; (**b**) 1912–1949; (**c**) 1949–1979; (**d**) 1980–2016

occupied by natural landscape. Traditional houses always avoid being adjacent to the main path, both physically and mentally. They are often located on the secondary path branching off from the main road, with an enclosed courtyard wall or a recessed space ahead serving as a transition to separate them from the public.

The modern street network began to be established in the village since the construction of Xiaolian Road in 1972. It is a preplanned road built by the county government to facilitate transportation in the mountainous eastern part of Taishun County. After that, several modern streets were built one by one from the 1980s to the 1990s. They constitute the major traffic structure that we can see now in Xiaqiao. In 2005, another big change in the transportation network happened: a new provincial road, No. 58, was built on the west side of Xiaqiao, which greatly facilitates the connection between the village and the outside world. After that, local residents could travel on the new road directly without transferring at other villages (Fig. 4.9).

The new street network changes the way in which the spaces of Xiaqiao are organized to adopt a transportation-oriented approach. It is not a locally originated creation. It is a product created by external society. The new streets break down the settlement's original spatial structure. The housing groups are disintegrated and reorganized.

As a totally artificially planned result, the new street system stands outside the natural development system of the traditional settlement. Instead of serving the daily circulation inside the village, it is more crucial to consider how to efficiently connect to the regional transportation network and the outside world. The surface of the street is also different from the traditional path system. Buildings are neatly lined along the street, forming the artificial elevation of the street.

The changes in the street network, as an external production inserted into the village, are a miniature of the conflict between tradition and modernity.

Fig. 4.9 Transformation of Street Network Connecting Xiaqiao to other places

4.2.2.2 Spatial Occupation

The spatial occupation in the village reflects the growing process of the built-up area in the village. It is measured by figure-ground diagrams that highlight the area occupied by buildings (Fig. 4.10).

The series of diagrams clearly reflects the expansion of the built-up area in the village. In the first three periods, the settlement expands gradually, and the rate of increase of new buildings is stable. But in the last market-economy period, the built-up area is enlarged so fast that most of the available land is occupied by new buildings.

The scale of construction was limited by the economic capability. In the first three periods, the limited capability of agriculture restricted the potential of settlement

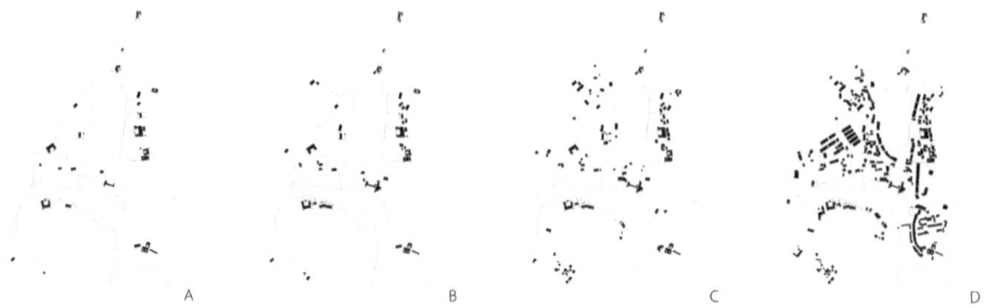

Fig. 4.10 Transformation of Built-up Occupation, (**a**) 1644–1911; (**b**) 1912–1949; (**c**) 1949–1979; (**d**) 1980–2016

development. But after the reform of the whole economic system in China, the rural society has developed greatly. With the increase in wealth, more construction projects were carried out.

The changing mode of land development is another reason contributing to the fast occupation of the built-up area in Xiaqiao. In recent years, instead of traditional self-construction conducted by the villagers themselves, the government has started to lead the general development of the village by learning from the real estate mode in the urban market. The scale of housing construction that has appeared successively in Xidong and Daligang is so large that it has never been seen in the village before.

4.2.2.3 Landscape

Landscape reflects how the settlement interacts with nature. It represents the domestication process aimed at controlling the natural environment. The landscape of Xiaqiao transforms from natural landscapes to artificial landscapes.

The natural landscape is the primary consideration in deciding whether people would settle down. The enclosing mountains and circuit rivers in Xiaqiao are favorite landscapes which attract immigrants. Chinese people have even developed an entire theory of Fengshui to interpret the natural world in terms of human spirituality. It gives more value to the natural environment. People also use significant landscapes as landmarks. For example, there are three big trees at the entrance to the road to remind passengers of the direction. In this stage, people are only equipped with the ability to select and visualize the beneficial natural landscape.

The transforming and artificializing of the natural landscape begin immediately after the primary stage that contains three steps. The three steps do not appear successively over time. There are overlapping periods between them. After transforming the whole natural setting into human-related settings, people try to improve it by artificial creations. To avoid the unfortunate landscape, people build constructions such as bridges and gateways, or rebuild their houses. It is the add-construction approach to artificialize nature.

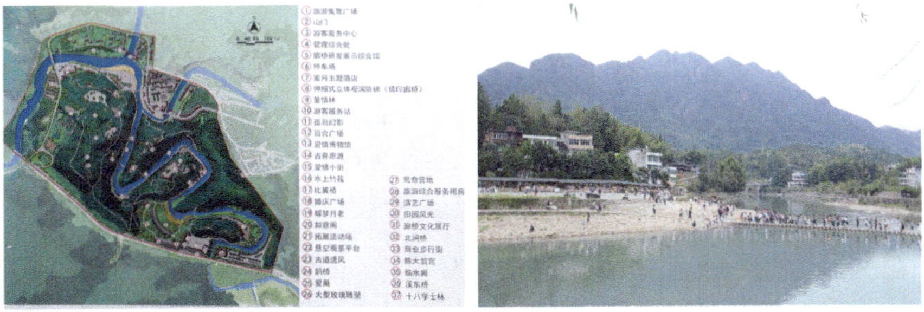

Fig. 4.11 Artificialization of landscape Left: The Planning of Langqiao Cultural Park by the Taishun County Government; Right: Current Landscape of Langqiao Cultural Park

Along with the enhancement of the ability to transform nature, people try to transform the original landscape. New farmlands are reclaimed from forests. Dams and water reservoirs are constructed to control the water flow in the river. New roads are built along the configuration of the mountains. It is the second step of artificializing nature. It starts in the Qing dynasty, but the ability to transform nature has been strengthened over time. Especially, the big infrastructures can only be built after the planned-economy period.

The third step of artificializing nature is to create landscapes, which emerged in the fourth period after 1980, especially in the recent decade. In this step, people try to create landscapes by construction. For example, the landscape along the river was built as Langqiao Cultural Park after 2015 (Fig. 4.11). The sandy areas of the river have been created. All the bank landscapes are created, such as sandy areas, vegetation, embankments, etc.

4.2.3 Maturity of Community

4.2.3.1 Density

The density of buildings and diversity of functions are used to describe the maturity of settlement communities. Building density serves to measure the intensity of construction within a settlement, while functional diversity is used to assess its capacity for self-sufficiency.

Building density reflects the proportion of built-up area within a settlement. Unlike spatial occupation criteria that focus on transforming spatial patterns, research on building density emphasizes quantitative changes. From the Qing Dynasty to the present, the building density of Xiaqiao has steadily increased (Table 4.3, Fig. 4.12). Population growth drove continuous construction to accommodate new residents. As farmlands were protected by both farmers and government policies, new buildings were prioritized around existing structures, resulting in higher density in older residential areas compared to newer ones.

Table 4.3 Transformation of construction density

Period		New construction area (m2)		Total construction area (m2)	Construction density	Diagram Ref.
The Qing Dynasty	1611–1911	/	/	14,641	0.012	A
Republican China	1912–1929	4409	6913	19,050	0.016	B
	1930–1949	2504		21,554	0.018	C
The Planned-economy Period	1949–1969	5295	7954	26,849	0.022	D
	1970–1979	2659		29,508	0.024	E
The Market-economy Period	1980–1989	2397	38,654	31,905	0.026	F
	1990–1999	10,583		42,488	0.035	G
	2000–2009	14,745		57,233	0.047	H
	2010–2016	10,929		68,162	0.056	I

In the first three periods, the building density of Xiaqiao shows steady growth. But after the 1980s, it increases rapidly. A large-scale construction planned and regulated by the government is inserted into the previously unexploited area. The new buildings are lined along the road or arranged in tight rows in the blank area (Fig. 4.13).

4.2.3.2 Programming

The extent of program diversity decides the way in which the community is involved in regional cooperation. The more diverse it is, the more interactive it is with the outside world. The types of programs that Xiaqiao contains experience a change from single-function to multi-function (Fig. 4.14). During the imperial period, Xiaqiao was the smallest residential unit. It equipped the basic programs that provided shelters for accommodation, farmlands for livelihood, and temples and ancestral halls for spiritual beliefs. From the end of the Qing to the Republican period, along with the rising status of traffic in Xiaqiao, fixed shops appeared around the Beijian Bridge and the Xidong Bridge.

In the planned economic PRC, the village was redefined and reorganized by the unit of communal brigades. Ideology imprints on space. A grand hall was built then. It is the biggest building in the village as a symbol of the political ideas of socialist society. A primary school was built next to the hall. It was the first public educational resource in the village that served the whole southeastern Taishun. In this period, due to political issues, the religious function was greatly weakened. All the temples and ancestral halls were used as storage or canteens for schools. As an independent community, the supply and marketing cooperative was built in the Xidong area, to provide non-staple food and groceries; a barber shop was also built in Xidong.

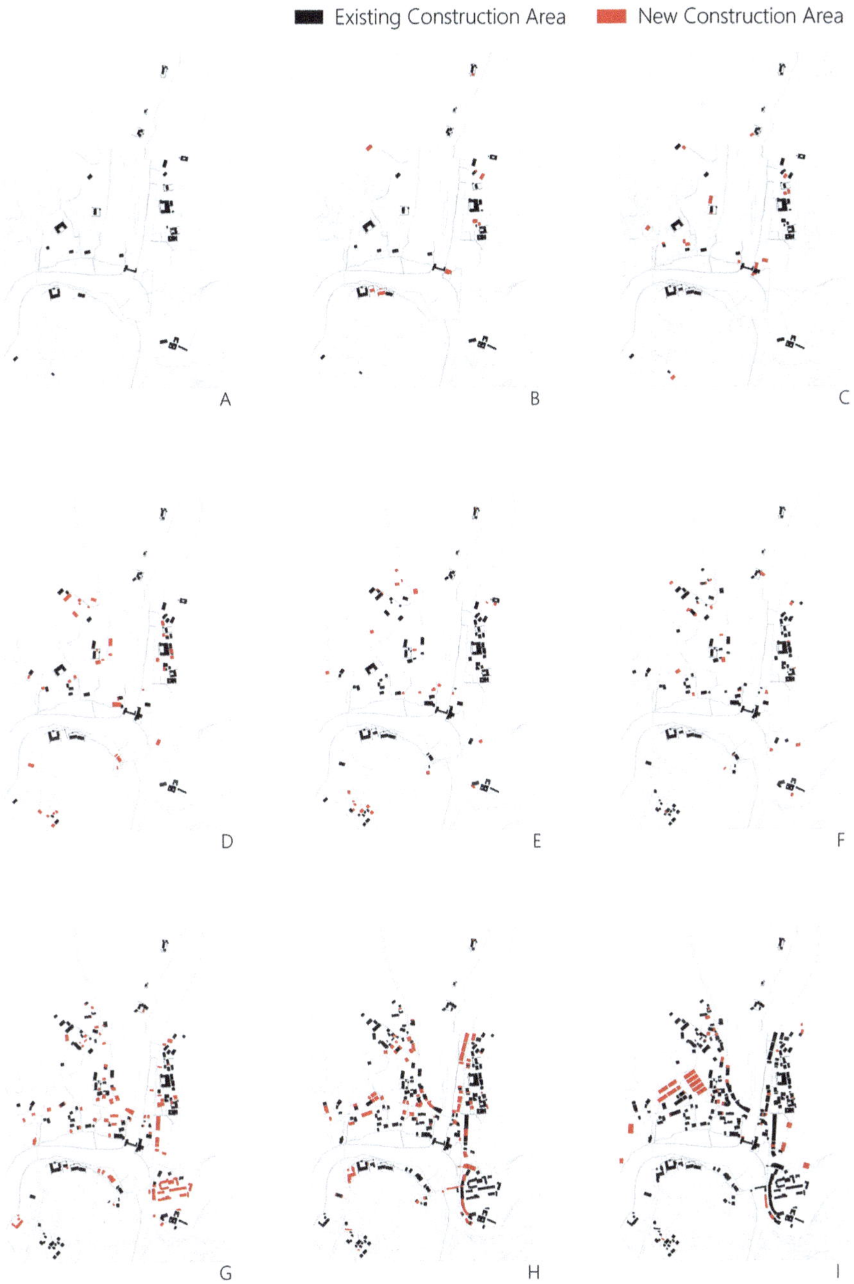

Fig. 4.12 New construction area in each period, (**a**) 1644–1911; (**b**) 1912–1929; (**c**) 1930–1948; (**d**) 1949–1969; (**e**) 1970–1979; (**f**) 1980–1989; (**g**) 1990–1999; (**h**) 2000–2009; (**i**) 2010–2016

Fig. 4.13 New buildings along the road

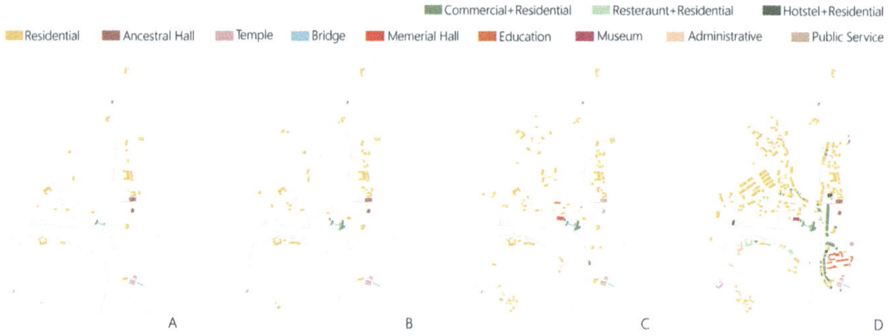

Fig. 4.14 Transformation of programming in Xiaqiao settlement, (**a**) 1644–1911; (**b**) 1912–1949; (**c**) 1949–1979; (**d**) 1980–2016

In the market-economy period, more functions enter the village. Economic constitution changes the community structure. Along the traffic lines, more shops and small businesses have opened to serve more people than just the local residents. New functional spaces are brought into Xiaqiao along with tourists. Nowadays, the village is not only used for residential purposes, but is also equipped with sightseeing and tourism value as a tourist destination. Souvenir shops, restaurants, hotels and hostels are common in the village since the establishment of Langqiao Cultural Park in 2007. They first appeared in the area surrounding the Beijian Bridge and the Dongxi Bridge, then spread into other parts of Xiaqiao along with the continuous development of tourism.

New identities as cultural heritage sites and tourist destinations are added to the original functions of Xiaqiao's buildings. To better showcase the village's unique bridge architecture, the grand hall, which is strategically located, was converted into a small museum. Urban visitors perceive traditional dwellings and bridges, which now serve as tourist attractions, as novel and interesting. Qing-dynasty courtyard houses, temples, and ancestral halls, designated as protected heritage sites, reflect historical value. Some religious buildings regained their original functions, while

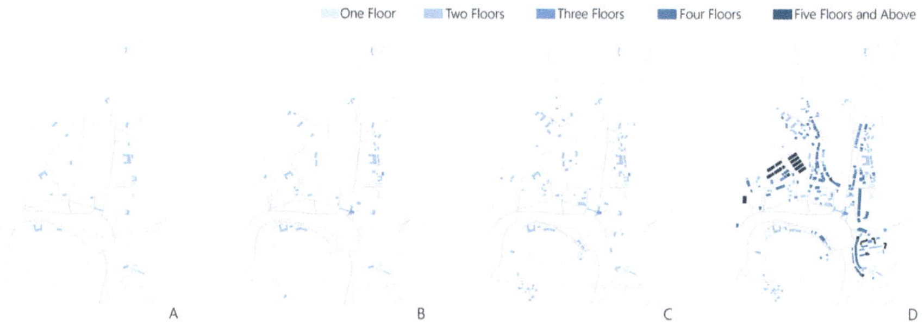

Fig. 4.15 Transformation of Building's Height in Xiaqiao Settlement, (**a**) 1644–1911; (**b**) 1912–1949; (**c**) 1949–1979; (**d**) 1980–2016

others were replaced by newly constructed temples and monasteries. For example, Chendaweng Temple was repurposed as a communal center.

Government control over village functions is evident in spatial transformations. The former primary school now houses the village committee office, while Taishun Second Middle School serves the southern region of the township.

4.2.4 Development of Architectural Technology

Five indicators—height, structure, material, construction period, and architectural form—are used to analyze architectural technological development (Fig. 4.15, 4.16, 4.17, 4.18 and 4.19). The evolutionary process is clearly visible in the diagrams: Qing-dynasty buildings were typically one to two stories high with wooden structures. During the Republican and planned-economy periods, mixed structures combining wood with stone or rammed earth emerged. After the 1980s, new technologies and materials became widespread, with building heights increasing to four to five stories, marking an overall construction upgrade in Xiaqiao.

4.2.5 Property of Clan Village

For a clan-based village, family lineage and land ownership are critical issues. Clan distribution reflects how different lineages controlled village territories. The village comprises five clans with six divisions: the Jinanlin Clan (including Houchi and

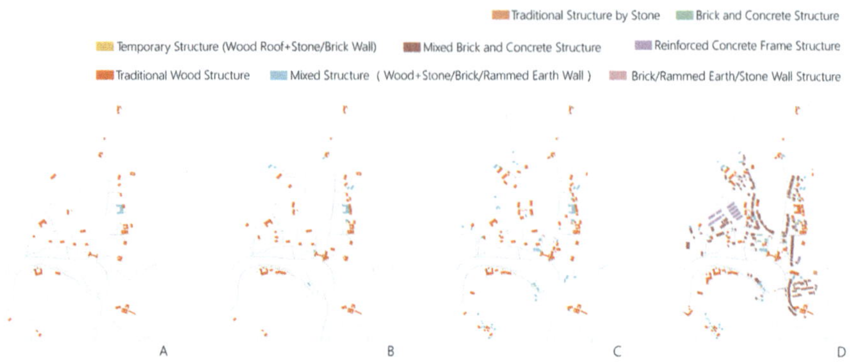

Fig. 4.16 Transformation of Structure in Xiaqiao Settlement, (**a**) 1644–1911; (**b**) 1912–1949; (**c**) 1949–1979; (**d**) 1980–2016

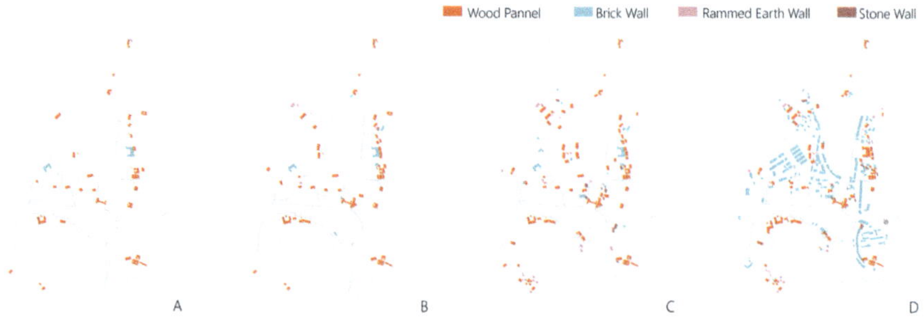

Fig. 4.17 Transformation of Building's Enveloping Material in Xiaqiao Settlement, (**a**) 1644–1911; (**b**) 1912–1949; (**c**) 1949–1979; (**d**) 1980–2016

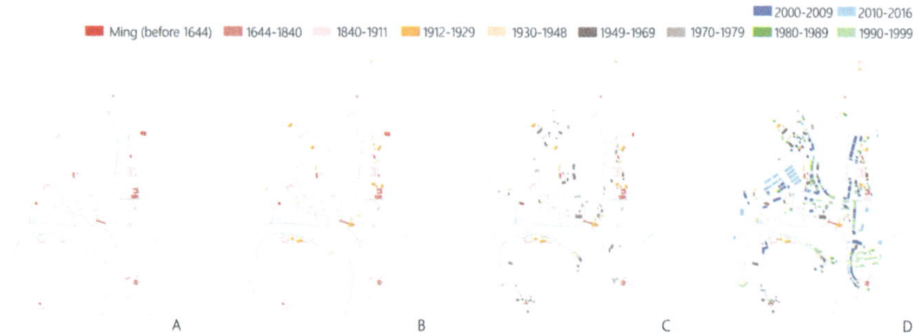

Fig. 4.18 Transformation of Existing Building's Built Period in Xiaqiao Settlement, (**a**) 1644–1911; (**b**) 1912–1949; (**c**) 1949–1979; (**d**) 1980–2016

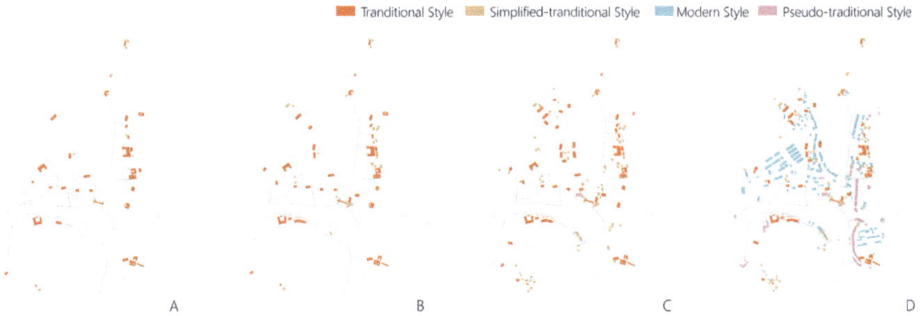

Fig. 4.19 Transformation of Building's Form in Xiaqiao Settlement, (**a**) 1644–1911; (**b**) 1912–1949; (**c**) 1949–1979; (**d**) 1980–2016

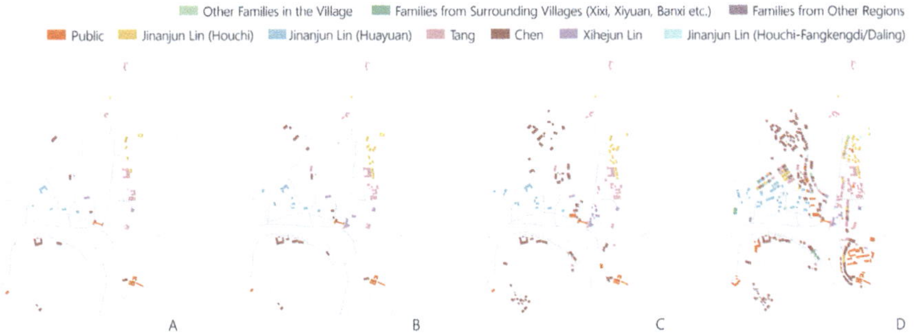

Fig. 4.20 Transformation of lineage distribution in Xiaqiao Settlement, (**a**) 1644–1911; (**b**) 1912–1949; (**c**) 1949–1979; (**d**) 1980–2016

Huayuan divisions), the Tang Clan, the Chen Clan, and the Xihe Lin Clan.[3] The governance model shifted from single-clan dominance to mixed residency (Fig. 4.20).

During the imperial period, families of the same lineage typically lived together due to cultural traditions and land ownership practices. Clan regulations prioritized intra-lineage land transactions to protect property, resulting in lineage-specific residential zones. However, clan fortunes fluctuated, leading to land transfers and shifts in residential rights.

During the Qing Dynasty, the village was, to some extent, self-governed by lineage groups, leading to significant territorial changes. Generally, the Houchi Lin division occupied the northern east bank of the North River; the Tang Clan lived south of the river and on the west bank; the Xihe Lin Clan resided east of Beijian Bridge; the Chen Clan lived west of the bridge and in the southern mountains; and the Huayuan Lin division settled on the north bank of the West River. While family

[3] Surname is the standard to differentiate different lineages. Different lineages have ancestors with different birth places. Thus, Jinanjun 济南郡 means their original ancestor was born in the place named Jinanjun, and Xihejun 西河郡 means their original ancestor was born in the place named Xihejun.

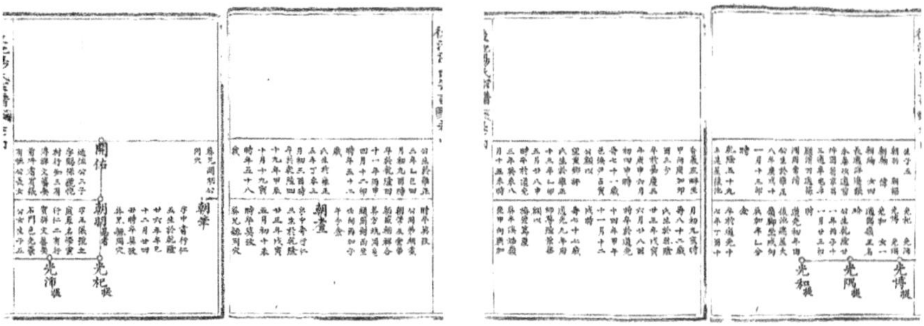

Fig. 4.21 Records of House Construction from the Family Archives of the Tang Family

zoning remained stable overall, competition in high-density areas caused territorial shifts among clans. As shown in the family archives of the Tang family in Fig. 4.21, members of the third generation of the Tang family had already begun building houses in the Houchi area, the population center of Xiaqiao Village (Fig. 4.21).

The territorial rivalry between the Houchi Lin division and the Tang Clan is notable. Initially, both riverbanks belonged to the Lin division, but the Tang Clan's rise in the mid-Qing period (eighteenth century, Appendix III) led to extensive land purchases, as recorded in Tang family land deeds (Fig. 4.22). The Tang Clan became the dominant landowner, controlling most of the west bank and half of the east bank, establishing the clan distribution pattern for subsequent periods.

In the Republican period, the pattern of family territory has not changed much in the original living quarters, but some new lands have been developed and filled with new residents. Both the North Mountain and the east bank of the South River have been resided by the Chen lineage.

From the planned-economy period, the single-lineage-controlled mode has been changed due to the direct involvement of the government in land distribution. Guided by the equalization ideology of communism,[4] the spare spaces of the big houses are re-allocated to the families who did not have enough living spaces although they are not the owners of the house. But the reallocation of the living spaces is still based on the consideration of lineage. Mostly, the new residents belong to the same lineage with the house's original owners but might be from different divisions.

However, after the 1980s, the residents' constitution of Xiaqiao became more complicated. Population mobility has been enhanced. Extraneous population has increased in the village. The lineage consideration has been entirely abandoned in the process of land distribution. The land transactions are all under the guidance of

[4] New PRC's land system can be traced back to the implementation of the China Land Law Outline in1947, which is the programmatic document to guide the rural land reform, and also laid the tone of the future land system changes. This document established the peasant private ownership. Article Three stipulates that the land "ownership is owned by the farmers." Article Six stipulates that "all the lands and the public lands of all landlords in the village, together with all the other lands in the village, according to the total population of the village, distribute equally regardless of men, women and children."

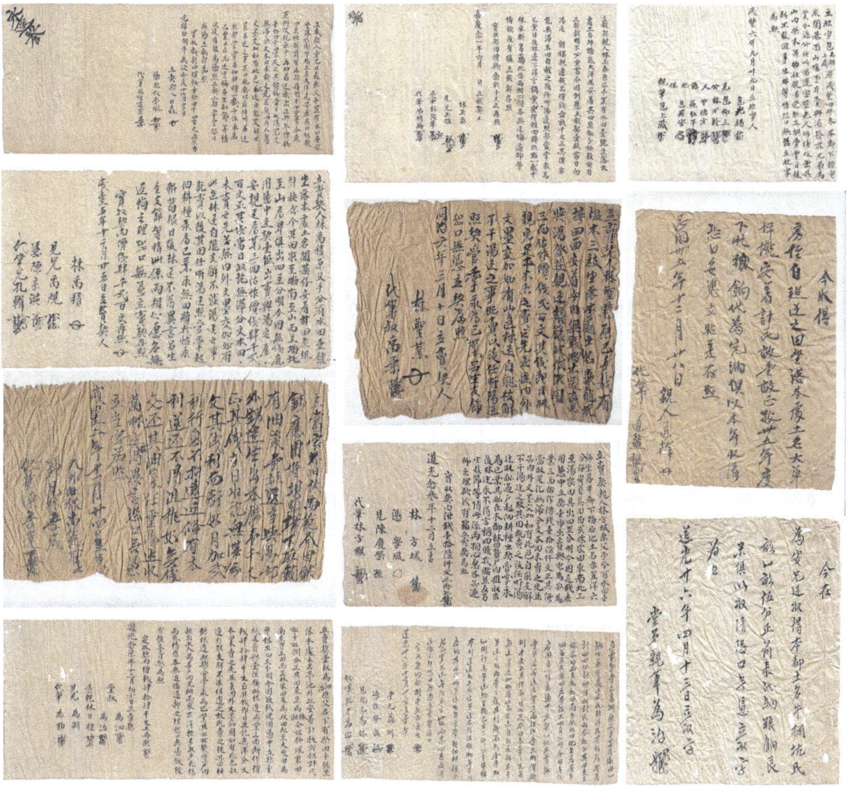

Fig. 4.22 Indeeds of land transaction of the Tang Lineage

the government. Two approaches are involved. One is to develop land into residential regions for the poor who lost their original houses in the disaster; the other is to develop land into commercial housing for anyone who would like to buy it. Both approaches break the original pattern that one living quarter is occupied by one major lineage. Instead of providing lands for one targeted lineage, it is for all purchasers, even those from other villages. De facto, the government encourages the concentration of population from peripheral villages into a central village. It develops the previously idle lands in the village into saleable lands for residence or business. It finally results in a mixed residential situation.

4.2.6 Comparative Study

4.2.6.1 The Impact of Lineage Control on Settlement Pattern: Combed Shape Layout, Guangdong

1. Description of Combed Shape Layout

The combed shape is the most common and typical layout of rural villages in the area near Guangzhou. The general shape of the settlement usually approximates a rectangle. All buildings are arranged orderly in rows and columns like a comb.

The appearance of a typical combed shape village is normalized with a fixed set of elements that contain a pond, square, ancestral hall, dwelling, and lane (Fig. 4.23). The pond is always located at the front of the village, practically for drainage, fish-raising, and irrigation, and mentally for acquiring fortune and wealth from the perspective of Fengshui (geomancy). Next to the pond is a spacious front square used as a grain-sunning ground [14]. All the village buildings are built after the square. Ancestral halls are in the first row, as the lead of common residences. Lanes are formed naturally between the neighboring two rows or two columns of buildings that connect to every corner in the village [15–16].

Zhonglou village (钟楼村) in Conghua (从化) and the Daqitou village (大旗头村) in Sanshui (三水) are two typical villages with a comb-type layout. The Zhonglou village has a spacious square at the front, followed by seven columns and seven rows of buildings (Fig. 4.24). The ancestral hall for Ouyang Renshan sits in the middle of the first row. Three-bay-two-gallery-style houses are located on both sides. The rear of the village has rows of auxiliary buildings with their doors facing the road. All the alleys are connected horizontally at the end by a back lane. Walls are built around the village, with two gates as the only openings on the front side.

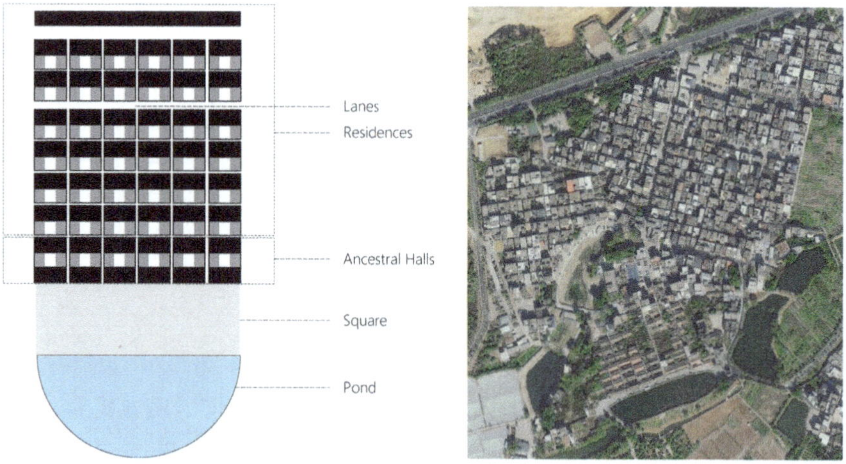

Fig. 4.23 Left: Typical layout of combed shape settlement; Right: Google map of Gangtou village

Fig. 4.24 Layout of the Zhonglou Village, Left: Figure-ground Plan; Right: Google map of Zhonglou Village

Another example, the Daqitou Village (Fig. 4.25) is planned integrally by the representative of the sixth generation, Zheng Shaozhong. As leaders in the village, Zheng Shaozhong and his relatives reside in the center of the village. All the buildings in the village are arranged in order. The first row is comprised of ancestral halls, temples, and big courtyard houses. Other dwellings are built behind them. Due to the long distance from north to south, an additional lane is set in the middle to facilitate the traffic. A spacious granite square was built before the buildings.

2. Impact by Lineage Control

Both natural and social conditions influence settlement patterns. Topographical features decide the general shape and size of settlements; policy, regional economy, local social structure etc. influence the built-up process. In this case, the clan system is the leading factor that forms the comb-shaped layout. Social structure has been projected in the morphological pattern, especially in the organization of public space.

The difference between the center layout and the comb layout is in inner spatial structures rather than in general shape. For the center layout, the ancestral hall is located in the center, which is surrounded by common dwellings. However, for the comb-shaped layout, ancestral halls are built in the first row of the village, facing water and leading the subsequent ordinary houses morphologically.

These two layouts have a temporal sequence of evolution in general. The center layout is an earlier type of the combed shape layout. It appeared in the period when only high-grade officials were legally allowed to build ancestral halls. Thus, the location of the ancestral Hall becomes the criterion for distinguishing different types of settlement forms. It decides the method by which the whole village is organized.

The difference in settlement pattern is the direct result of responding to the needs of the clan system in Guangdong. According to the research of Friedman [17–18], the clan system in China can be divided into two extreme states by internal fission, A

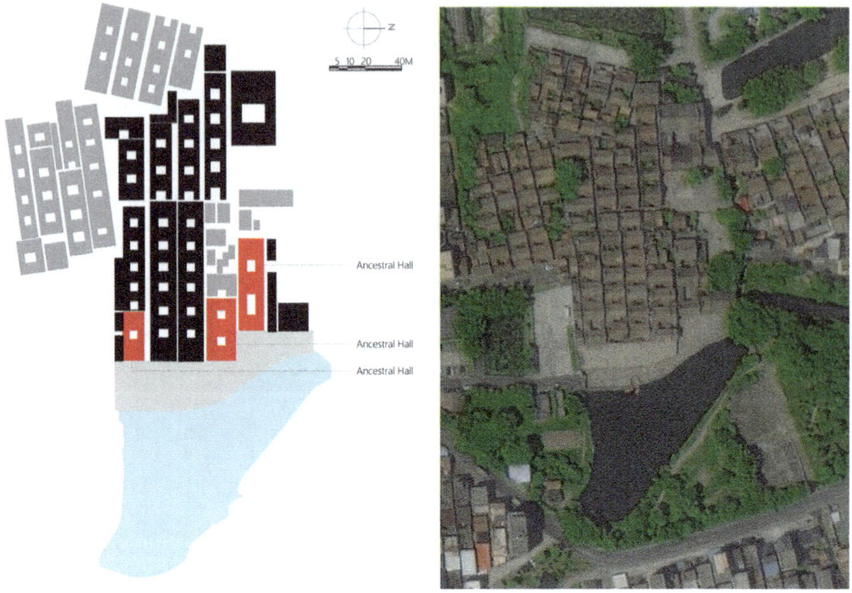

Fig. 4.25 Layout of the Daqitou Village, Left: Figure-ground Plan; Right: Google map of the Daqitou Village

and Z. The former belongs to the smaller and poorer clan communities without sub-level fractions. Compared to the poorer fractions, the richer ones are better able to pay for building new shrines and progenitor settings. In general, at the establishment period of a rural settlement when few generations are included, the lineage belongs to 'Type A' internal division. Only one ancestral hall was built in the center. It corresponds to the center type of settlement pattern. However, the lineage belongs to 'Type Z' internal division, with several powerful divisions co-existing in the village, after experiencing a longer period of development. More than one ancestral hall is built when the general lineage is divided. Morphologically, the increasing number of ancestral halls increases the need to change the original central pattern to emphasize the importance of each division.

The prerequisite of the combed shape layout in Guangdong relates to social changes. Friedman notes that the clan phenomenon is closely related to the current official ideology. The legalization of ordinaries' ancestral halls begins from the publication of two policies in the Ming dynasty, "Da Li Yi (Great Ceremony, 大礼议)" and the subsequent one, "Tui En Ling (Push Grace, 推恩令)".[5] It made an important impact on the clan system of Canton. After that, ancestral halls were built all over

[5] Tui En Ling 推恩令, Da Li Yi and Tui En Ling are both released in the 15th year of the Jiajing period, the Ming dynasty. They are policies about the rites to worship family ancestors. They allow all the common peoples in the official ancestors instead of only the officials in previous periods.

Guangdong. It changed the pattern that an absolute center, as the ancestral hall of the only high-level official family, was formed in rural settlement. To build ancestral halls, a correspondingly great amount of wealth is needed to support the construction. This requirement can be satisfied by the economic development of Guangdong during the same period. In the Jiajing (嘉靖) period of the Ming dynasty, when the construction of the clan system was at its peak in Guangzhou, a great scale of new land cultivation began. It brings sufficient wealth to establish new settlements and build new ancestral halls. In the newly established settlements, several ancestral halls could be built by different divisions of a lineage that were parallel to each other on the same level.

4.2.6.2 The Impact of Defense Requirement on Settlement Pattern: Ring-shaped Layout, Fujian

1. Description of Ring-shape Layout

The ring-shaped layout (Fig. 4.26) is a unique layout of rural villages in the Hakka region containing Raoping in Guangdong and Zhao'an and Yunxiao in Fujian. The shape is usually circular or semi-circular, varying from the layout of the Tulou (Rammed Earth Building, 土楼) complex. The settlement with a ring-shaped layout has obvious centripetal expansibility and clear geometric spatial structures. The center is always an ancestral hall. All the dwellings are arranged around the center in an orderly way, round and round like water ripples [19]. They are built gradually with the growth of population.

Centripetal and enclosing are two important morphological features of the ring-shaped settlement. The multiple rings continue to spread out as a wave in which the outermost circle forms a boundary. It serves to differentiate the local lineage from outsiders and suggests the field of the settlement. While highlighting the significance of ancestor sacrifice, it also reflects the blood relationship between different divisions

Fig. 4.26 Typical layout of Ring Shape Settlement

Fig. 4.27 Layout of Ring Shape Settlement in Raoping, Guangdong Province

inside the lineage. The center is undoubtedly the focal point of composition that is always occupied by the ancestral hall (Fig. 4.27).

The multiple rings of the settlement are not finished in one go. In the case of Yunxiangzhai (Fig. 4.28), initially, only two to three rings of dwellings were built. More rings are added along with the development of the village, and finally the settlement forms five and a half rings. The spatial form of ring-shaped settlement reflects the imprint of social cohesion in traditional clan society.

2. Impact by Defense Requirement

The formation of the ring-shaped layout is a complicated process in response to the defense requirement. The chaos of the unstable social environment is the direct cause. The powerful lineage system is the planning and organization of the whole construction project. The historical inheritance of living customs is the cultural basis for living settlements with a circular shape.

First, the insecure social environment raises the requirement to build a concentrated and closed settlement. This concentrated and closed settlement is a defense feature of the ring-shape settlement. Then, an inwardly closed settlement layout mainly refers to a settlement with security and defense features. In history, the region where the ring-shape settlements are distributed continuously experiences wars, clan fights, and other threats from bandits and pirates. But the military power of the official government is inadequate. How to protect themselves is significant when building a new settlement. Thus, the ring-shape layout settlement is generated with an enclosed spatial feature to protect the residents' lives and properties [20].

Fig. 4.28 Layout of Ring Shape Settlement in Zhao'an, Fujian Province

Officials also encouraged small and scattered settlements to merge into a big one with walls and fortresses for protection. According to the records in "Annals of Guangdong" of the Jiajing period. In the 38th year, pirates attacked Raoping from Fujian. The government "commanded people who lived in scattered settlements to move to big villages. Villages need to be protected by their male adult residents. Residents who live near the city must move into the city" [21]. This kind of policy led to a significant change in the settlement pattern after the mid-Ming dynasty, with a marked militarization process, or fortification process, of settlements into big fortress-style villages with military walls and a concentrated layout pattern. Most of the current existing cases of ring-shape villages in the junction region of Guangdong and Fujian are formed after this period.

Secondly, the ring-shaped layout has a close relation with the lineage system as well. The construction of a well-planned village needs to be controlled by a strict and powerful social organization and financed by the accumulation of wealth of the entire lineage. Houses' size, orientation, and arrangement pattern are arranged by lineage. The centralized spatial pattern in which all dwellings face the ancestral hall reflects a rigorous hierarchy of lineage by emphasizing the status of ancestor worship.

The people who organize the construction, in general, should keep a good relationship with the government. This type of layout not only offers a sense of security for the residents who live there, but also reflects both a manifestation of orthodoxy in the government and a symbol of wealth and strength in the countryside.

The formation of the ring-shaped layout is a cultural genesis influenced by living customs. Many existing structures in contemporary society are, in fact, formed historically under specific natural and social conditions. It is the result of the accumulation of history and culture. But after being created, it acquires its own independence. It is interesting to see how the features persist.

4.2.6.3 The Impact of Geographical and Social Limitation on Settlement Pattern: Dispersed Layout, Taishun

1. Description of Dispersed Layout

The dispersed configuration is the most common and typical layout of rural villages in the mountainous area of Wenzhou, especially Taishun. The configuration of settlements depends on the specific topography. The shape is irregular and homogenized, without an obvious focus on spatial pattern. All buildings are distributed, scattered, and fitted to the terrain, utilizing the available land while minimizing the occupation of farmland. The buildings are built independently without overall planning.

2. Impact by Geographical and Social Limitation

The southern Zhejiang is occupied by crowded mountains. The limitation of topography impacts the formation of a dispersed layout. It fits the natural conditions closely. Due to limited land and an irregular geographical environment, the dispersed layout of a village usually does not have a fixed and regular shape.

According to the terrain, the dispersed layout can be divided into four groups: basin, riverbank, the intersection of several rivers, and mountain slope. The four types of layout have different configurations. In the basin layout, all buildings are dispersed in the basin, showing a slug-like pattern which is almost an irregular polygon or roughly square; in the riverbank layout, all buildings are distributed along the river, banded villages: village houses near the water flow along the riverside on one or both sides; at the intersection of rivers, all buildings are distributed around the intersection area; in the mountain slope layout, all buildings are arranged along the contour lines of the mountain (Fig. 4.29).

Dispersed layout is also determined by the basic relationship between people (houses and houses), as well as between people and society (private houses and public buildings). There are two social relations existing in rural China: family and clan. Family refers to the blood relatives within five Fu, while lineage, as a broader concept, refers to the relations of the five Fu. The full-range order of social relationships, including immediate and collateral relatives, is called "the Pattern of Difference Sequence" [22].[6] The unique social structure in rural Chinese society impacts on the living mode of rural residents.

The lineage system in Taishun is simple but complete. As symbols of a mature lineage, ancestral halls, family archives, and lineage farmland are commonly seen in every rural settlement in Taishun. Usually, one village is settled by one lineage or co-settled by two or more lineages.

[6] The pattern of difference sequence: (or "difference pattern") The term is proposed by Chinese scholar Fei Xiaotong to describe the interpersonal relationship of the Chinese traditional society. He described the Chinese tradition Social interpersonal relationship, as the configuration of ripples when the stone goes into water. The ripple formed by layered of circles from the the center that represents the person him or herself, to outside that represent others. The distance between the circle and the center is to distinguish how close the relationship is between the people and others.

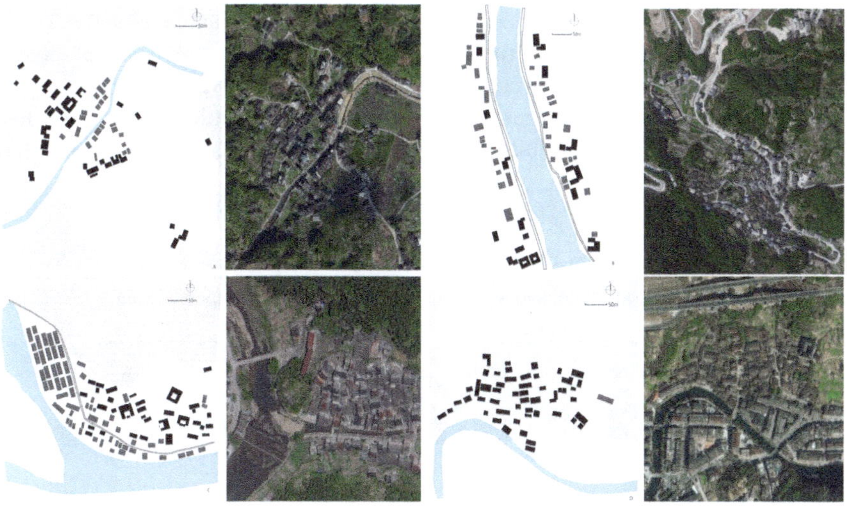

Fig. 4.29 Different types of dispersed layout in Taishun (**a**) the Baifuyan village; (**b**) the Xuelin village; (**c**) the Shiyang Town; (**d**) the Kucun village

However, the lineage system is not as strong as that of Guangdong and Fujian due to the undeveloped economy. As the quantity and quality of farmlands are both limited by the geographical environment, the extent of economic development is greatly reduced in rural society where agriculture is the major source of productivity. Thus, a developed and powerful lineage that can control a large amount of resources cannot be established.

Compared to planning the village entirely by the centralized power of the lineage, the dispersed settlements in Taishun are formed gradually through the independent construction of houses by each family. The management of the village is flat, so the social hierarchy is not as obvious as that in Guangdong and Fujian. Every big house is built separately from others. It is the projection of the village's social structure onto the settlement pattern. Thus, how the society is organized is a significant factor contributing to the formation of the layout.

4.2.6.4 Comparison and Conclusion

Settlement patterns provide specific historical information. It is obvious that the middle of the Ming dynasty is an important time slot for the formation of rural settlements in southeast China. New policies were published in the Jiajing period and continued to influence the following hundreds of years.

Different settlements reflect different social structures (Tables 4.4 and 4.5). The combed shaped layout of rural settlements near Guangzhou shows that the control of a powerful lineage that grasps most local resources can dominate regional development. The ring-shaped layout of rural settlements in the conjunction area of Guangdong and Fujian provinces reflects how the settlements transformed in response to the urgent needs of social changes. The dispersed layout of settlements in Taishun is the projection of the limited natural and social environment onto the spatial structure .

Table 4.4 Comparison of three settlement types (Part I)

Analytical Parameters	Combed shape layout	Ring-shape layout	Disperse layout
Shape	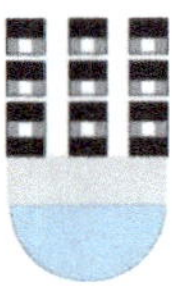		
Location	Guangdong	Fujian	Wenzhou
Dwellings			
	Rows and columns	Surround type	Scattered, no fixed type
Ancestral hall			
	At the front row	In the center	No fixed location
Planned	Yes	Yes	No
Orientation	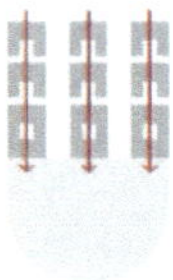	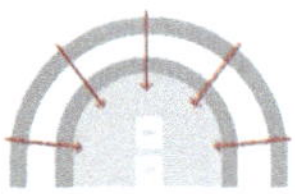	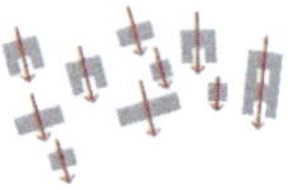
	Same direction	Face the center	No fixed orientation
Fengshui	Consider the whole settlement	Consider the whole settlement	Consider single building
Size of settlement	Big	Middle	Small
Model	Compact	Compact	Disperse
Substitution of settlement	Clearly apart	Clearly apart	Mix together

Table 4.5 Comparison of three settlement types (part II)

Village type Analytical Parameters	Combed shape layout	Ring-shape layout	Disperse layout
Major Contradiction	Resource control	Family Security Protection	Access to living resources
Economy	Exploration of Shatian	Extreme contradiction between land and people	Limited farmland
	Developed	Normal	Undeveloped
	Da Li Yi(The Great Ritual Debate)	Qian Jie Jin Hai (Maritime Ban); The threat of pirate and riots	The official development of the Taishun County
Period	From the middle of the Ming dynasty	From the middle of the Ming dynasty	From the middle of the Ming dynasty
Lineage System	Complicate	Complete, unified	Complete
	Powerful	Powerful	Common
Geography	Coastal plain	Mountainous area	Mountainous area

In general, the pattern of settlement reflects the lives of local ethnic groups. All the cultures are the product of human activities. The formation of a specific settlement pattern has experienced a complicated process in history that is affected by various factors. The spatial structure has not been created in one go. Instead, it has experienced a dynamic development process over hundreds of years in different specific spatial and temporal contexts. Thus, the spatial structure of settlement is a dynamic historical process, a result of accumulation by multiple layers of historical periods.

4.2.7 Summary of the Second Level

The settlement level is the second of seven levels. The objective of this study is to examine the transformation of Xiaqiao village as an integral unit. The settlement pattern is the result of the synthesis of information from multiple layers of historical periods. A thorough analysis of the five groups of analytical factors reveals a clear progression in the development of the entire village, spanning from the Qing dynasty to the planned-economy period. Nevertheless, beginning in the 1980s, there was a marked shift in settlement patterns, characterized by significant alterations in both the quantity and quality of buildings. This transformation was marked by the construction of numerous new structures, leading to a shift in spatial patterns from a focus on natural environments to one that was more oriented toward traffic and industry.

At the settlement level, the pattern of settlement reflects the life of the local ethnic group. All culture is the product of human activity. The formation of a particular settlement pattern is the result of a multifaceted historical process that is influenced by various simulations. The spatial structure is not created in a single instance; rather, it undergoes a dynamic development process over the course of hundreds of years, encompassing multiple specific spatial and temporal configurations. Consequently, the spatial configuration of settlements is a dynamic process in history, resulting from the accumulation of multiple layers representing distinct historical periods.

4.3 Level 3: Sub-settlement Level

The third level is the sub-settlement level to describe the spatial pattern of different zones inside the Xiaqiao Village. It tries to explain the transformation of the rural built environment by breaking down the settlement into different naturally grown residential groups and studying them comparatively. It continues the discussion on settlement issues, but instead of stopping at the general observation of the entire settlement, it focuses on different parts of Xiaqiao to explain the inner tensions in depth.

At this level, a comparative study of seven subsettlements in the village has been conducted. A series of morphological diagrams is drawn according to different variables. They can show the development sequence of the settlement and different development features under different situations. Then case studies are conducted to explain the specific types of development modes stimulated by different conditions.

The seven subsettlements are defined by residential zones that formed spontaneously in history. They are called 'natural villages', a concept different from 'administrative villages'. An administrative village refers to a larger area that belongs to the same administrative unit defined by the government. A natural village refers to a small field equal to a settled unit that was established by the primary settlers at the beginning.

The Xiaqiao Village is an administrative village that constitutes several residential zones. In the county archive of the Qing dynasty *Fengjianglu* [10] (Records of Field Dividing, 分疆录), the Xiaqiao Village is recorded separately by different residential areas such as Huayuan, Houchi, Nanshan, Qiaotou, etc. According to the records in the 1984 Taishun Dimingzhi (Archives of Taishun's Toponymy, 泰顺地名志), the whole Xiaqiao Village is reorganized as the Xiqiao Production Brigade (下桥公社) during the communization period of 1958 [23]. Twelve production teams are included, which are Shanhou (山后), Shenshan (沈山), Fangkengdi (芳坑底), Kengwei (坑尾), Tangqiangang (堂前岗), Qiaotou (桥头), Yangtou (垟头), Beishanqiao (北山桥), Houchi (后池), Huayuan (花园), Xidong (溪东), and Nanshan (南山)[23]. Among them, the first five are peripheral residential areas that are far away from central Xiaqiao, have little relation to the development of the other parts of the village, and have been discarded now. Thus, in this chapter, only the last seven subsettlements are discussed (Fig. 4.30).

The analysis is conducted from three perspectives, which are the adaptability to the natural environment, the establishment of residential community, and the strength of development. There are 13 criteria involved, which are terminology, shape, spacing,

Fig. 4.30 Zoning of Seven Sub-settlement in Xiaqiao

Fengshui, programming, lineage occupation, built-up occupation, density, height, structure, material, built period, and architectural form (Table 4.6).

4.3.1 Adaptability to Natural Environment

Subsettlements are residential areas settled by villagers spontaneously in accordance with the specific geographical conditions. They reflect how the artificial space adapts to nature.

4.3.1.1 Terminology

The terminology used to name subsettlements created by the locals clearly describes the specific topography. People can locate them easily by reading the names. Qiaotou (entrance of the bridge, 桥头) is the area next to the Beijian Bridge; Yangtou (entrance

Table 4.6 Analyzing element of Subsettlement Pattern

Analyzing Objects	Criteria	Evaluating Factors
Adaptability to natural environment	Terminology	Relate to natural landscape; relate to artificial landmark
	Shape	Centralized / De-centralized; Linear / conglomerated
	Spacing	Exist / disappear; Clear defined / gradually transitioned/mixed
	Fengshui	Mountain surrounding / water agriculture landscape
Community establishment	Program	Residential; shop; educational; temple; bridge; storage; service; restaurant; hostel and hotel; mixed (residential and commercial);
	Lineage occupation	The Houchi Lin lineage; the Huayuan Lin linage; the Tang lineage; the Chen lineage; the Qiaotou Lin lineage; families from adjacent areas; families from other areas in Taishun; families form other area outside Taishun
Strength of development	Built-up occupied	Built-up area / open space
	Density	High/medial/low
	Height	One floor / two floors / three floors / four floors / five floors / six floors
	Structure	Traditional wood structure; mixed structure I (wood frame structure plus stone/brick/rammed earth load Bearing Wall); mixed brick structure; brick and concrete structure; mixed structure II (brick walls plus concrete frame structure)
	Enveloping's material	Wood panel / Brick Wall / Stone Wall / Rammed Earth Wall
	Built period	The End of Ming Dynasty; The Early of Qing Dynasty; The Late of Qing Dynasty; The Republican China Period (1912–1948); The Planned-economy Period (1949–1979); The Market-economy Period (1980–2016)
	Architectural form	Traditional style / modern style / simplified traditional style / pseudo-traditional style

of the farmland, 垟头) is the area next to the large farmlands; Beishanqiao (bridge of North Mountain, 北山桥) is the area on the slope of North Mountain; Houchi (behind the pond, 后池) is the area near the pond of Tang's ancestral hall; Huayuan (garden, 花园) is the area that originally has a garden built by the Lin lineage; Xidong (east side of the river, 溪东) is the area on the east bank of the Dongxi River, and Nanshan (south mountain, 南山) is the area on the slope of South Mountain.

4.3.1.2 Shape

Shapes form in accordance with specific topography. Shape is used to describe the distribution pattern of buildings. A linear shape is formed when buildings are built along the river, such as Houchi, Huayuan, and Xidong, or along the contour of the mountain slope, such as Nanshan and Yangtou. A terraced shape emerges when buildings are constructed on different levels of a mountain, as in the case of Beishanqiao. An agglomerate shape is formed when buildings are clustered in a relatively plain area, such as Qiaotou. When the geographical conditions change, the configuration of the newly formed built-up area changes accordingly (Table 4.7).

4.3.1.3 Spacing

The spacing between each pair of sub-settlements has shortened over time. During premodern times, every sub-settlement was rather independent, at a great distance from one another. But along with continuous construction activities, the built-up area enlarged greatly. This led to the merging of the neighboring sub-settlements (Fig. 4.31).

4.3.1.4 Fengshui

Fengshui is a comprehensive result of culture, spirit, and living experience. It was the primary factor that residents needed to consider when they first settled down in premodern China. Xiaqiao has not been built entirely under the guidance of unified planning. Instead, it grows naturally by the unit of each sub-settlement. Thus, it is more meaningful to discuss Fengshui at this level. All seven sub-settlements face the river and sit against the mountains, which is in accordance with the standard of good Fengshui (Fig. 4.32).

4.3.2 Community Establishment

A sub-settlement is usually an independent living community in the village. It reflects how the community is established.

4.3.2.1 Programming

Programming reflects the main function of the community. During premodern times, most of the sub-settlements were single-functional residential areas, except for Qiaotou, where a few commercial functions emerged. The diversity of programming in the village has emerged in the past few decades. With booming tourism,

Table 4.7 Relation between Sub-settlement's Shape and Location

Shape	Location	Name	Diagram	Diagram
			Location	Shape
Linear shape	Along the river on the Bank	Houchi		
		Huayuan	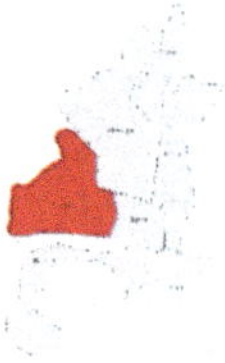	
		Xidong		
	Along the contour on mountain slope	Nanshan		
		Yangtou		

(continued)

Table 4.7 (continued)

Shape	Location	Name	Diagram	Diagram
			Location	Shape
Terraced shape	Different levels of mountain	Beishanqiao		
Agglomerate shape	Plain area	Qiaotou		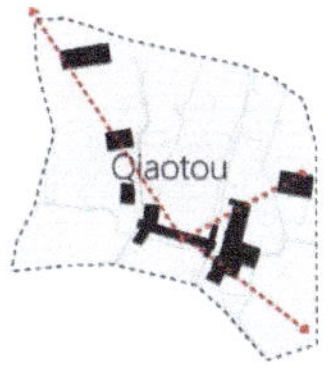

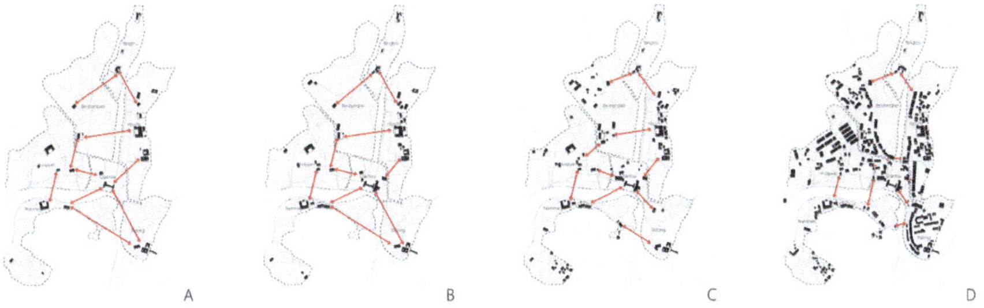

Fig. 4.31 Transformation of spacing between adjacent sub-settlements, (**a**) 1644–1911; (**b**) 1912–1949; (**c**) 1949–1979; (**d**) 1980–2016

more citizens come to the village to see the traditional dwellings and bridges, and many service functions have grown accordingly. The differentiation of programs in different sub-settlements began at almost the same time. Qiaotou, Nanshan, and Xidong have incorporated more commercial functions compared to the remaining four sub-settlements, which are still mainly residential. Among them, Xidong became a commercial street serving tourists, and Qiaotou became the central area of a cultural park that has been planned by the county government since 2005 (Fig. 4.33).

4.3.2.2 Clan Occupation

The analysis of lineage occupation at the sub-settlement level clearly shows the real controlling areas of each lineage in premodern times: Houchi is co-occupied by the Tang lineage and the Houchi Lin fraction of the Jinan Lin lineage; Qiaotou is

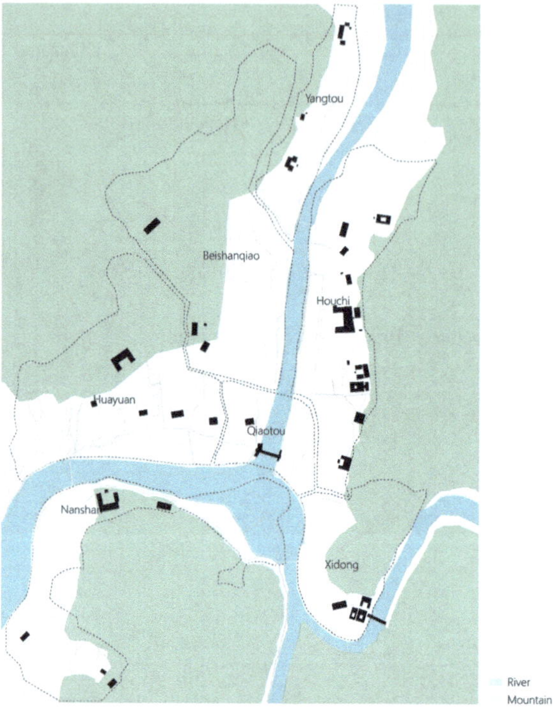

Fig. 4.32 Fengshui pattern of the Seven Sub-settlements

co-occupied by the Qiaotou Lin lineage, the Tang lineage, and the Chen lineage; Huayuan is occupied by the Huayuan Lin fraction of the Jinan Lin lineage; Yangtou is occupied by the Tang lineage, but the Lin lineage's ancestral hall was once located there; Nanshan, Beishanqiao, and Xidong are mainly occupied by the Chen lineage, but families from the Tang lineage also built houses in some areas. But now, in Xidong, Xiaqiao, and Huayuan, people from different lineages, even from other villages, are co-living together (Fig. 4.34).

4.3.3 Strength of Development

The development of the seven sub-settlements follows different sequences. In the same period, different sub-settlements are at different development stages. For the same sub-settlement, its development proceeds at different rates in different periods.

4.3.3.1 Built-up Occupation

The diagram of built-up occupation shows the enlargement of built-up areas in different sub-settlements over the four socio-economic periods (Fig. 4.35).

A: Transformation of Sub-settlement's Programming: Beishanqiao

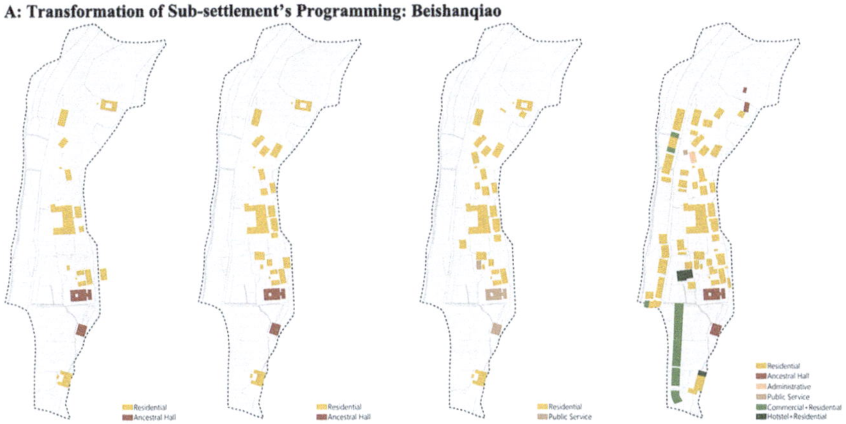

B: Transformation of Sub-settlement's Programming: Houchi

C: Transformation of Sub-settlement's Programming: Huyuan

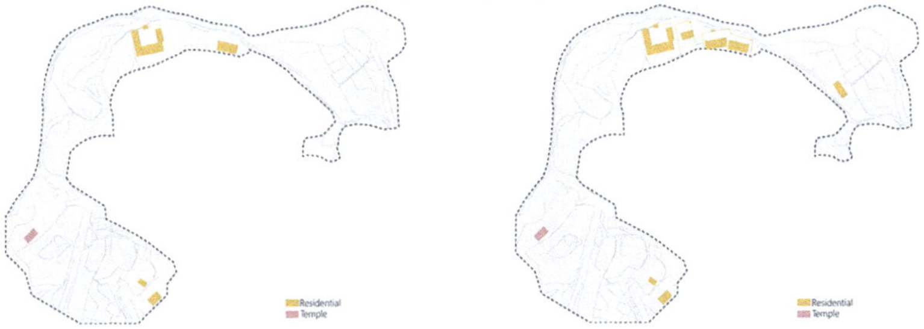

Fig. 4.33 Transformation of Sub-settlement's Programming in Beishanqiao, Houchi, Huayuan, Nanshan, Qiaotou, Xidong, Yangtou. (**a**) Transformation of Sub-settlement's Programming: Beishanqiao; (**b**) Transformation of Sub-settlement's Programming: Houchi; (**c**) Transformation of Sub-settlement's Programming: Huayuan; (**d**) Transformation of Sub-settlement's Programming: Nanshan; (**e**) Transformation of Sub-settlement's Programming: Qiaotou; (**f**) Transformation of Sub-settlement's Programming: Xidong; (**g**) Transformation of Sub-settlement's Programming: Yangtou

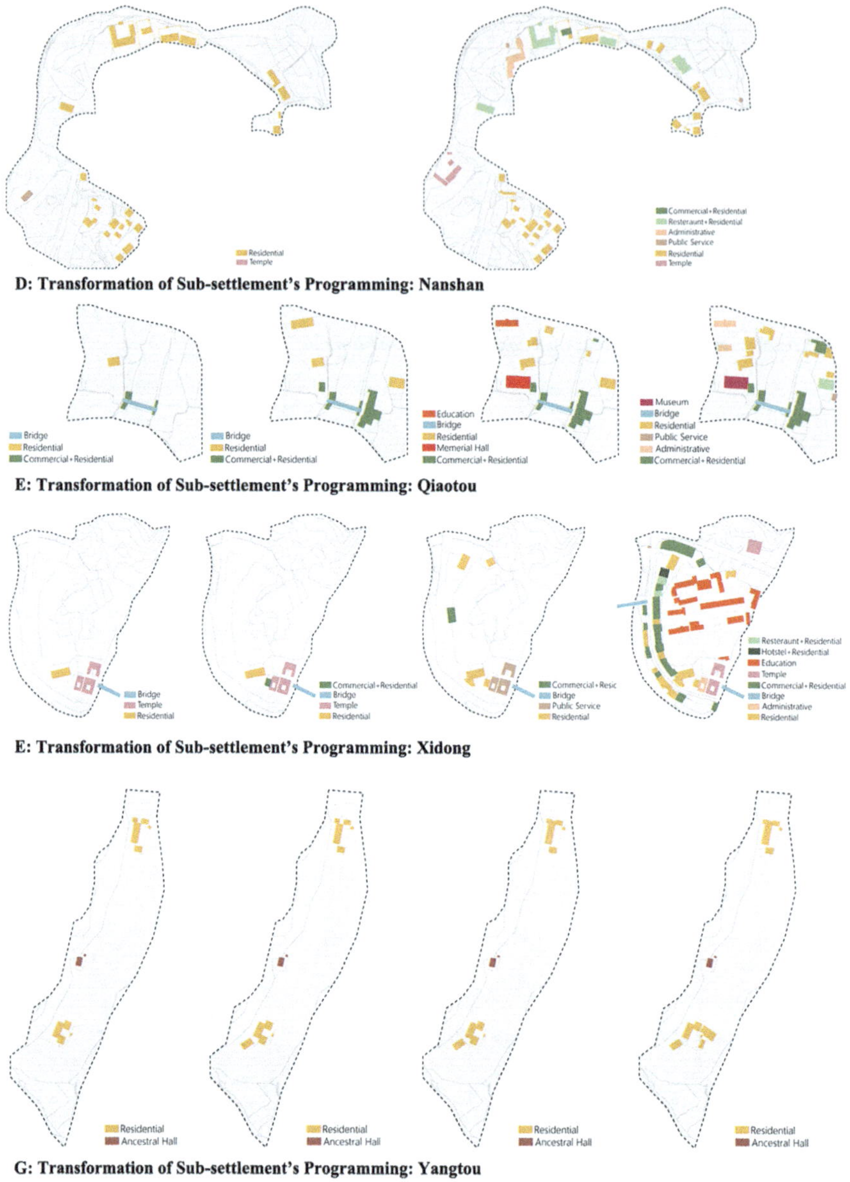

Fig. 4.33 (continued)

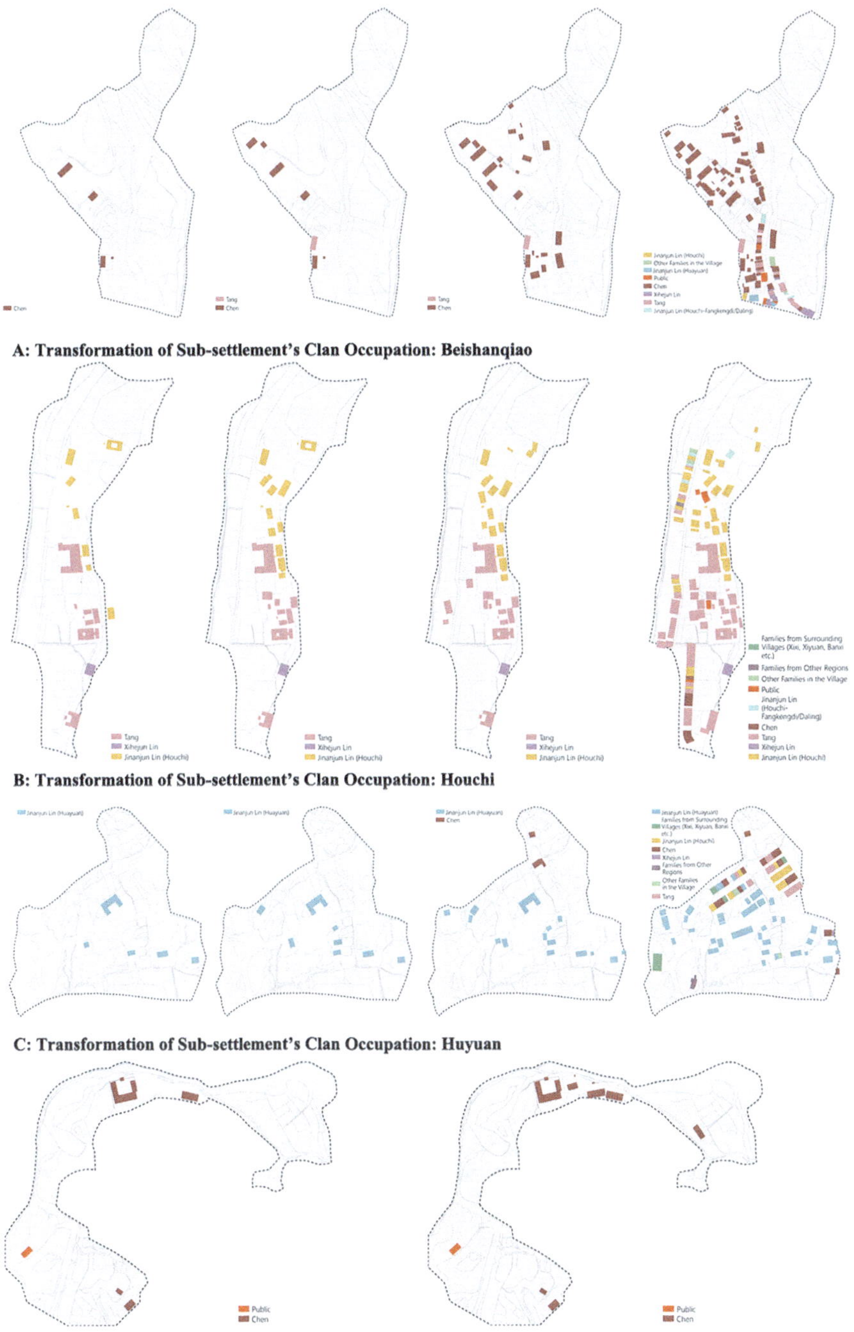

Fig. 4.34 Transformation of Sub-settlement's Clan Occupation in Beishanqiao, Houchi, Huayuan, Nanshan, Qiaotou, Xidong, Yangtou. (**a**) Transformation of Sub-settlement's Clan Occupation: Beishanqiao; (**b**) Transformation of Sub-settlement's Clan Occupation: Houchi; (**c**) Transformation of Sub-settlement's Clan Occupation: Huayuan; (**d**) Transformation of Sub-settlement's Clan Occupation: Nanshan; (**e**) Transformation of Sub-settlement's Clan Occupation: Qiaotou; (**f**) Transformation of Sub-settlement's Clan Occupation: Xidong; (**g**) Transformation of Sub-settlement's Clan Occupation: Yangtou

Fig. 4.34 (continued)

A: Transformation of Sub-settlement's Built-up Area: Beishanqiao

B: Transformation of Sub-settlement's Built-up Area: Houchi

C: Transformation of Sub-settlement's Built-up Area: Huyuan

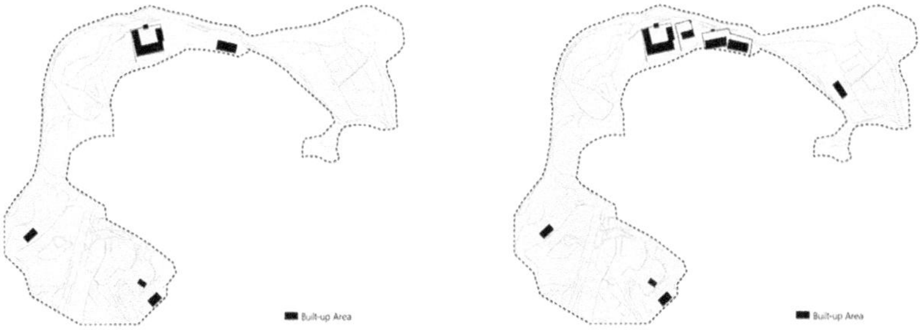

Fig. 4.35 Transformation of Sub-settlement's Built-up Area in Beishanqiao, Houchi, Huayuan, Nanshan, Qiaotou, Xidong, Yangtou. (**a**) Transformation of Sub-settlement's Built-up Area: Beishanqiao; (**b**) Transformation of Sub-settlement's Built-up Area: Houchi; (**c**) Transformation of Sub-settlement's Built-up Area: Huayuan; (**d**) Transformation of Sub-settlement's Built-up Area: Nanshan; (**e**) Transformation of Sub-settlement's Built-up Area: Qiaotou; (**f**) Transformation of Sub-settlement's Built-up Area: Xidong; (**g**) Transformation of Sub-settlement's Built-up Area: Yangtou

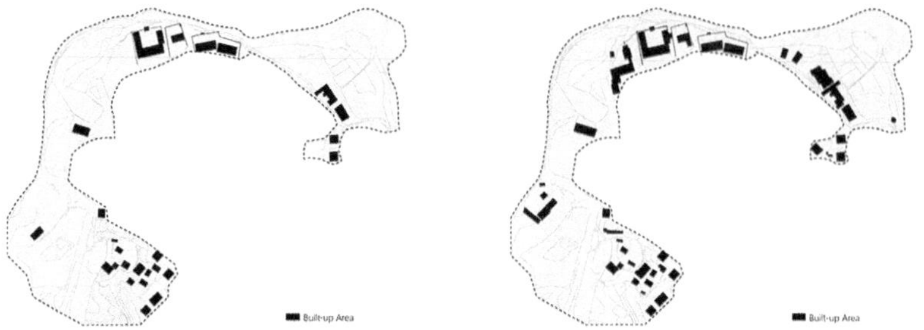

D: Transformation of Sub-settlement's Built-up Area: Nanshan

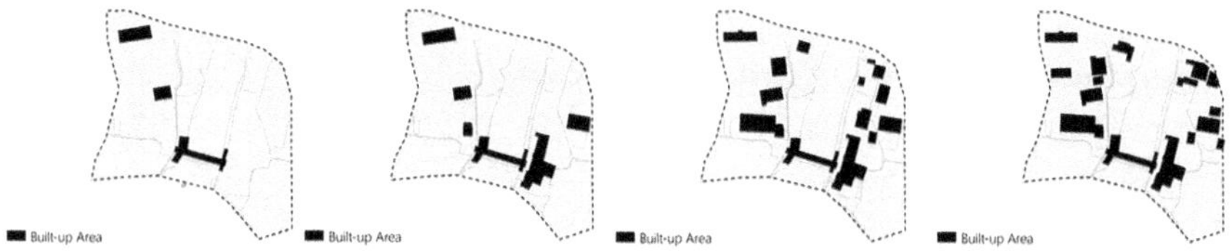

E: Transformation of Sub-settlement's Built-up Area: Qiaotou

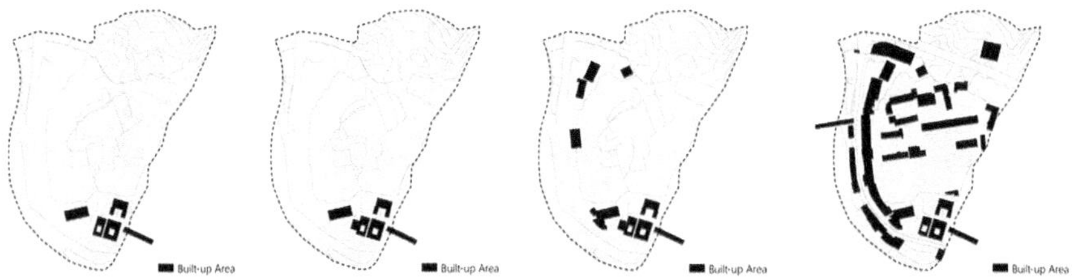

E: Transformation of Sub-settlement's Built-up Area: Xidong

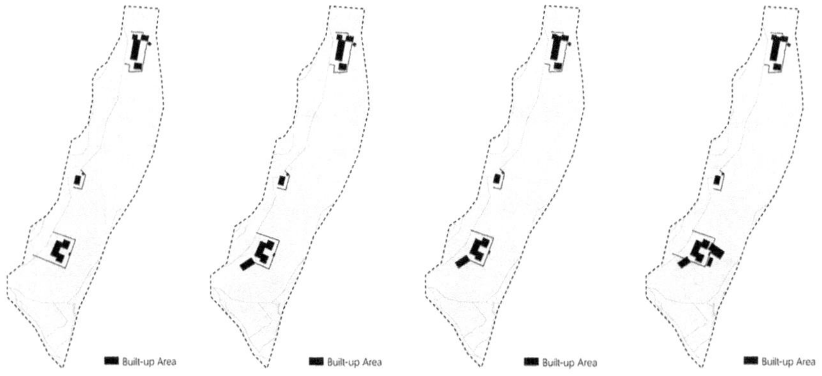

G: Transformation of Sub-settlement's Built-up Area: Yangtou

Fig. 4.35 (continued)

4.3.3.2 Density

The changing density clearly shows the strength of development over time among different settlements, although buildings are continuously being built everywhere, Xidong, Huayuan, and Houchi are the areas with the fastest rate of increase in building density (Fig. 4.36, Table 4.8).

The five other architectural features (height, structure, material, built period, and architectural form) show growth in construction from both perspectives of quality and quantity, and they can be seen in Appendix IV.

4.3.4 Features of Sub-settlement's Transformation

The seven sub-settlements play different roles in the village, and their roles keep changing over different periods. Several features emerge in the development process.

4.3.4.1 Asynchronism of Sub-settlements' Development

Although the social relation is much simpler and the area it occupies is much smaller in rural communities than in urban ones, the development of rural settlements is more complex than ever thought in the common view. Disparity does exist in the development of the built environment among different sub-settlements. They follow different paces and develop to different extents at the same time. For the same sub-settlement, the development rate is also not constant.

From the perspective of temporal sequence (Fig. 4.37) which different sub-settlements' development follows, Houchi, Huayuan and Nanshan are the earliest sub-settlements established in Xiaqiao before the Qing dynasty. Yangtou and Qiaotou were first developed in the middle of the Qing dynasty. Beishanqiao and Xidong were established only from the end of the Qing dynasty.

After being first settled, the seven sub-settlements have developed along different routes. Houchi has been the major residential area in the village for all the four periods discussed in this research where the building density is higher than other regions. A lot of new houses are built after 2005 under the governmental planning for the villagers who lost their house in the flood. The development of Houchi in modern times also benefits from its location next to the major intersection of roads.

Beishanqiao and Xidong are the other two sub-settlements where development in recent decades contributes to the construction of new roads to some extent. The greatest enlargement of them is after the 1980s along with the construction of the Xiaolian road. But for Xidong, the lands along the road were redistributed in the 1970s, and the large-scale construction occurred due to the establishment of Langqiao cultural park by the county government after 2005. It brings opportunity to the development of Xidong, as the main commercial street serving for tourists. Comparatively, Beishanqiao has little direct relations with the tourism booming in Xiaqiao these

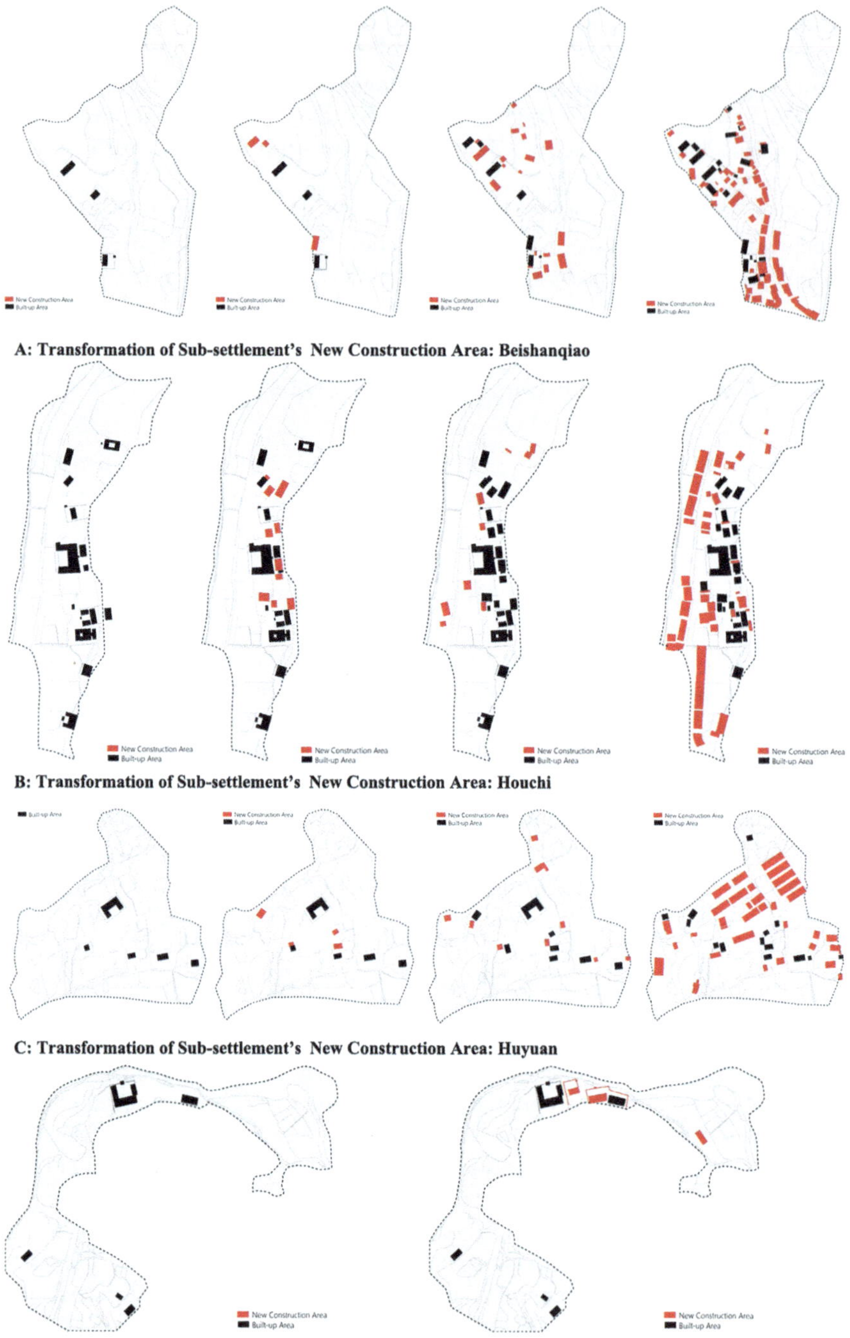

Fig. 4.36 New construction area in each period in Beishanqiao, Houchi, Huayuan, Nanshan, Qiaotou, Xidong, Yangtou
(**a**) (1–4): New Construction Area in Each Period: Beishanqiao; (**b**) (1–4): New Construction Area in Each Period: Houchi; (**c**) (1–4): New Construction Area in Each Period: Huayuan; (**d**) (1–4): New Construction Area in Each Period: Nanshan; (**e**) (1–4): New Construction Area in Each Period: Qiaotou; (**f**) (1–4): New Construction Area in Each Period: Xidong; (**g**) (1–4): New Construction Area in Each Period: Yangtou

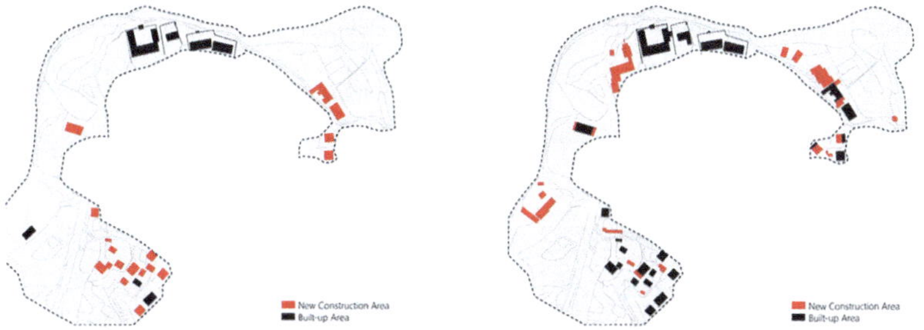

D: Transformation of Sub-settlement's New Construction Area: Nanshan

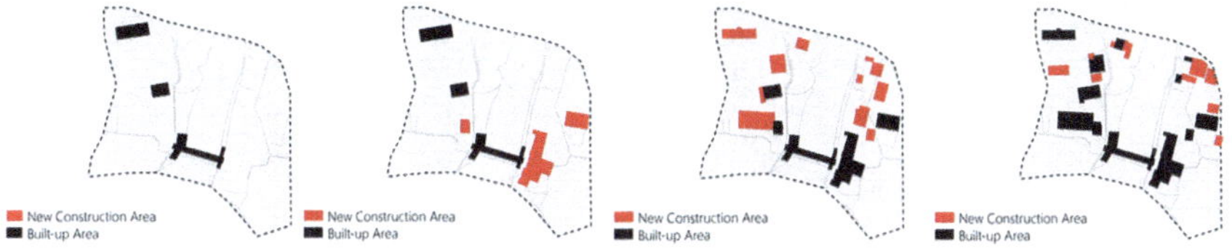

E: Transformation of Sub-settlement's New Construction Area: Qiaotou

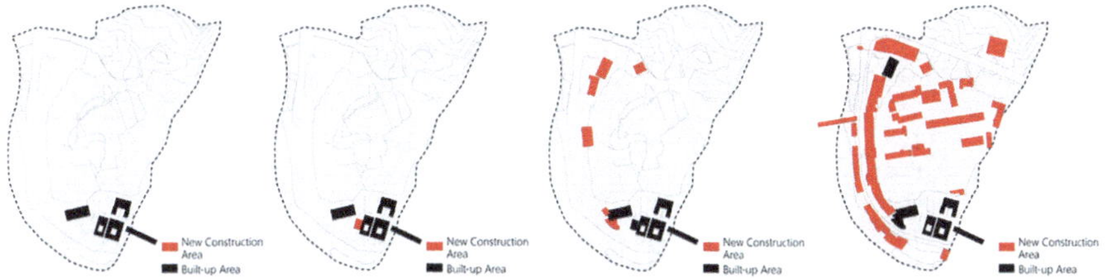

E: Transformation of Sub-settlement's New Construction Area: Xidong

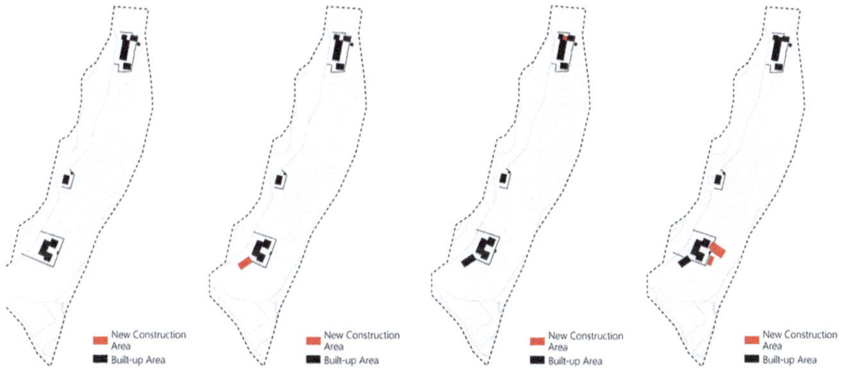

G: Transformation of Sub-settlement's New Construction Area: Yangtou

Fig. 4.36 (continued)

Table 4.8 Transformation of Sub-settlement's Construction Density

Sub-settlement	Period		New construction area (m2)	Total construction area (m2)	Construction density	Diagram Ref.
Beishanqiao	The Qing Dynasty	1611–1911	/	814	0.008	A-1
	Republican China	1912–1949	924	1738	0.014	A-2
	The Planned-economy Period	1949–1979	2848	4586	0.038	A-3
	The Market-economy Period	1980–2016	8700	13,286	0.11	A-4
Houchi	The Qing Dynasty	1611–1911	/	4560	0.057	B-1
	Republican China	1912–1949	1417	5977	0.074	B-2
	The Planned-economy Period	1949–1979	869	6846	0.085	B-3
	The Market-economy Period	1980–2016	7537	14,383	0.179	B-4
Huayuan	The Qing Dynasty	1611–1911	/	1561	0.016	C-1
	Republican China	1912–1949	1020	2581	0.026	C-2
	The Planned-economy Period	1949–1979	853	3434	0.035	C-3
	The Market-economy Period	1980–2016	9824	13,258	0.133	C-4
Nanshan	The Qing Dynasty	1611–1911	/	2024	0.026	D-1
	Republican China	1912–1949	2705	4729	0.061	D-2
	The Planned-economy Period	1949–1979	2139	6868	0.089	D-3
	The Market-economy Period	1980–2016	2466	9334	0.120	D-4
Qiaotou	The Qing Dynasty	1611–1911	/	1059	0.039	E-1
	Republican China	1912–1949	1173	2232	0.082	E-2
	The Planned-economy Period	1949–1979	1487	3719	0.137	E-3
	The Market-economy Period	1980–2016	446	4165	0.153	E-4
Xidong	The Qing Dynasty	1611–1911	/	1328	0.026	F-1
	Republican China	1912–1949	112	1440	0.029	F-2
	The Planned-economy Period	1949–1979	1055	2495	0.049	F-3
	The Market-economy Period	1980–2016	7027	9522	0.189	F-4

(continued)

Table 4.8 (continued)

Sub-settlement	Period		New construction area (m2)	Total construction area (m2)	Construction density	Diagram Ref.
Yangtou	The Qing Dynasty	1611–1911	/	1363	0.038	G-1
	Republican China	1912–1949	102	1465	0.041	G-2
	The Planned-economy Period	1949–1979	54	1519	0.043	G-3
	The Market-economy Period	1980–2016	263	1782	0.050	G-4

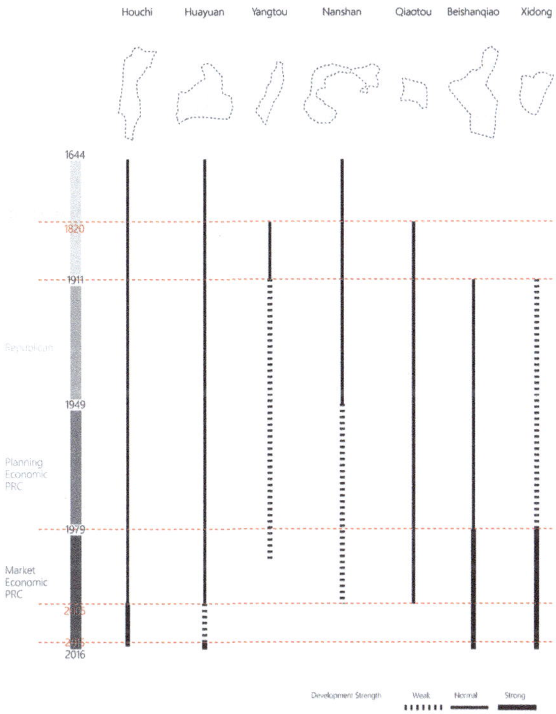

Fig. 4.37 Temporal sequence of Sub-settlement's development in Xiaqiao. (Drawn by the Author, 2017)

years. All the houses are only for residential. But new houses are always built along the road instead of on the mountain slope after the completion of Xiaolian road.

Nanshandi (南山底) is almost abandoned now due to its inconvenient location, with the residents moving to the area close to central Xiaqiao. Nanshan and Yangtou developed more slowly after the Republican period due to geographical limitations. Nanshan was mostly occupied by the Chen lineage. At first, they lived in Nanshandi, a peripheral region on the opposite side of Nanshan mountain that did not face the village's center directly. In the late Qing dynasty, they started to move to the side directly facing Huayuan. Several big courtyard houses were built during that period. But due to limited land for housing, fewer houses were built in Nanshan compared to other areas. Yangtou has hardly grown after the Qing dynasty, due to its inconvenient location at the corner of the village.

Huayuan and Qiaotou are two areas where development has been greatly constrained by the government in recent decades. During the Qing dynasty, Huanyang, along with Houchi, was the center of Xiaqiao, where more people settled down. Qiaotou is also prosperous because of the Beijian bridge, as both the regional spiritual symbol and traffic hub. In the 1970s, the great assembly hall was built next to the bridge as well. But after 2005, all the new construction halted in these two sub-settlements, due to the establishment of the newly established 'cultural park'. All the areas of Qiaotou and the major areas in Huayuan are covered by the park. The government wants to replace the original residences by other facilities related to tourism.

Generally, the development situation of sub-settlements is decided by the changing geographical and social factors. However, during the self-growing period in premodern China, including the Qing dynasty, the Republican period, and most of the planned-economy period, the differences in development among different sub-settlements were not obvious. The huge polarization happened after 1980 when the intervention and control from the government were intensified. Some areas have been constructed at an unprecedented speed. The large-scale constructions overturned the original pattern of settlement quickly. Besides the fast increase in the built-up area, measures to control folk construction were also implemented to unify the buildings under official planning.

4.3.4.2 Ordered or Disorder: Rural Metabolism in Sub-settlement's Growth

The metabolism of the village refers to how sub-settlements are created and upgraded, and how the old and destroyed buildings are replaced by the new ones. The metabolism process of sub-settlements in Xiaqiao experiences three stages, self-construction, governmental intervention, and overall planning by government (Fig. 4.38, Table 4.9).

1. Stage 1st: Self-construction

Self-construction is the major approach to build sub-settlement in premodern China. It contains two periods generally in this case, the Qing dynasty and the Republican period. The crucial features of rural society in the Republican period are inherited from the Qing dynasty directly. It cannot establish an effective management in village by the government directly, due to continuous wars. Thus, in this stage, almost all the construction activities in the village are initiated, organized, and conducted by the villagers themselves hardly with official supervision.

All the considerations related housing, the will to build, the preparation to building, and the organization of building process, in this stage all base on personal requirement, the real needs for living. Villagers decide whether it is needed to build new

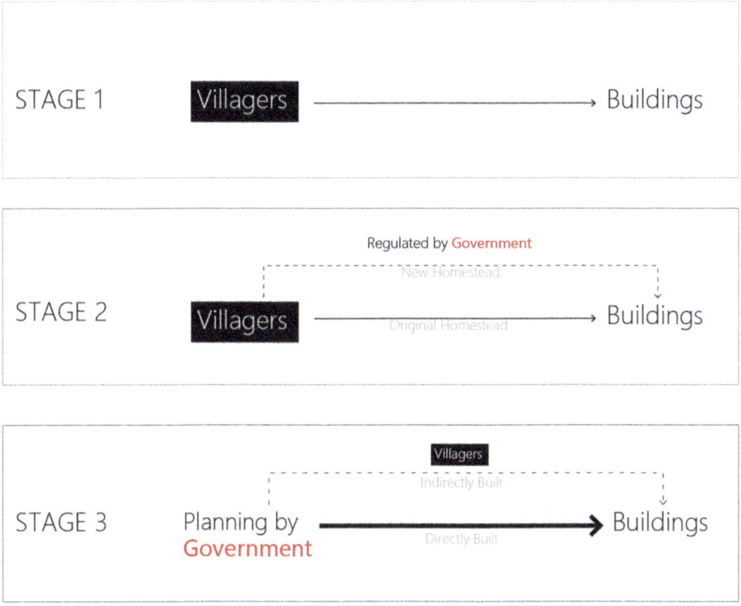

Fig. 4.38 Diagram of three stages

house, or whether it is affordable for them to build new house, by themselves, based on their own living level. After the decision of building new houses, the selection of site and transaction of lands are conducted by their own as well. Villagers looked for suitable builders, decide the style of house and the size of courtyard, according to their own preferences.

All the building behaviors are independent cases without inner relations at the level of overall planning of the settlement. New constructions only need to consider the situations they are specifically facing, instead of considering how to contribute to the overall situation the entire village faces. Thus, the houses are built one by one or upgraded one by one. It seems to be a disordered model for the metabolic process of each sub-settlement. However, the basic needs for housing decide the model's inner logic and order clearly.

2. Stage 2nd: Official Interven

After the establishment of PRC, the direct management of villages by government are strengthened. The folk construction behaviors need to be controlled under the supervision of government. The lands cannot be transacted freely among common villagers. Beside of the original homestead inherited from precedents, the new-added homesteads need to be approved by the government. Otherwise, the construction, even for basic housing, is treated as illegal behaviors.

However, beside of the restriction on approving new site for housing, the construction activities are still self-organized by the villagers their own. The villagers who

Table 4.9 Features of three development stages of sub-settlements

Stage	Period	Features	Diagram	
Self-construction	Qing Dynasty 1644–1912	Responding to the needs of daily life		
	Republican China 1912–1949			
Governmental intervention	The Planned-economy Period 1949–1979	Responding to the needs of daily life, but under the regulation of government	1. A storage of Beishanqiao production brigade was built. 2. A workshop as the community's tea factory was built.	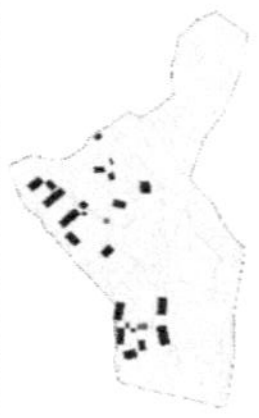

(continued)

Table 4.9 (continued)

Stage	Period	Features	Diagram	
			1. The big hous Houqiangdi is redistributed to several families as residence. 2. A storage of Houchi production brigade was built. 3. The ancestral halls were re-functioned as storages.	
			1. A primary school was built. 2. A great assemble hall as symbolic architecture of communal ideology was built.	
			1. A cooperative shop was built. 2. The temples were re-functioned as storages and public canteens.	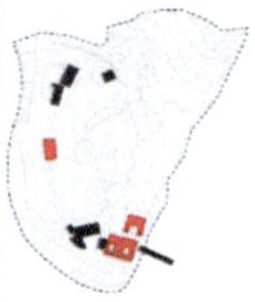

(continued)

Table 4.9 (continued)

Stage	Period	Features	Diagram	
Overall planning by government	The Market-economy Period 1980–2016	Controlled by the planning from government directly	1. The houses were built along the development of road under the overall planning of the county's road network. 2. Public service buildings were built, like fire station, and elderly culture center.	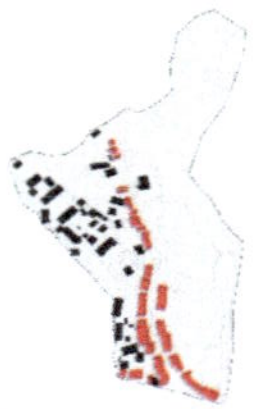
			1. The houses along the road were built as the settled houses for people who lost their houses in the flood of 2005. 2. The big house was protected as cultural heritage currently. 3. The ancestral halls return to their original ancestor worship function. 4. The houses along the road were decorated by the government as façade upgrading project.	
			1. The rowed houses were built as the settled houses for people replaced from the protection area of Langqiao Cultural Park. 2. A hostel was opened with the approve by government.	

(continued)

Table 4.9 (continued)

Stage	Period	Features	Diagram	
			1. The buildings as the entrance of Langqiao Cultural Park were built. 2. The big house were protected as historical dwellings, meanwhile were changed into restaurant and hostels for tourism with the approve by government. 3. The new Linshui temple was built.	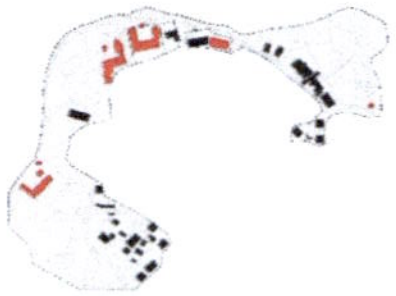
			1. The administrative building as the office of village committee were built. 2. The great hall was changed into museum for tourism by government. 3. The whole Qiaotou area was defined as the protection area of Langqiao Cultural Park by government where new constructions by villagers themselves were forbidden.	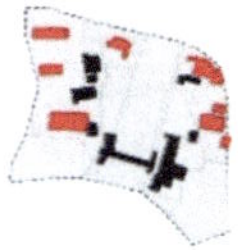
			1. The houses along the street was defined as commercial street for tourism by government. 2. The houses along the road were decorated by the government as façade upgrading project. 3. The Chendaweng temple was used as the administrative office of heritage protection. 4. The new middle school was built. 5. The Linshui temple was protected as cultural heritage currently.	

Fig. 4.39 Great Assemble Hall (**a**) Appearance (**b**) Indoor Space (**c**) Location

are willing to build new house need to apply for new homestead by transferring their own farmlands for the government. After obtaining approve, the rest building process are all grasped by their own. The semi-self model is a transitioning statue from self-construction to governmental planning.

The new type of constructing buildings that are initiated and organized by the government appears. Government started to build buildings for public use directly. The completion of this kind of building always gathers the forces from all the villagers. It is a new way to upgrade the rural built environment in Xiaqiao. A great assemble hall (Fig. 4.39) and a group of school buildings are built under the supervision of government directly by the villagers.

3. Stage 3rd: Governmental Planning

After the economic reformation in the 1980s, the direct management and control from the government over rural settlements has been greatly enhanced according to a series of land policies. The government is strictly involved in the land-controlling process. The approval of new homesteads is under strict supervision from multi-leveled governments. Household lands of individual families are expropriated and bought by the government due to various reasons relating to constructions of new infrastructures, such as new roads, new dams, new bridges, etc. It causes the appearance of land-requisitioned peasants. For this group of villagers, even the semi-self-construction model of housing becomes impossible due to no land can be transferred into homestead.

In recent decades, especially after the establishment of Langqiao Cultural Park in 2005, the government has played an enforced role in regulating construction behaviors. New buildings are strictly forbidden in the areas located inside the park. The government applies various methods to convert residential lands into public use by encouraging residents to move to outside areas. Besides land control, the government even enacts new regulations to regulate the envelope and façade of the buildings in the field that might influence the overall appearance of the cultural park. Buildings are restricted to a height below six floors in order to control the visual corridor. For the major street serving tourists, the decoration project is made uniformly in a style that imitates traditional style.

All the constructions are conducted under overall planning by the government. The upgrading of Xiaqiao is conducted under a purposeful sequence. Due to the strong power of the government, large-scale land development becomes possible. It is another change in the metabolism approach.

In the first two stages, the settlement metabolism targeted only specific buildings. Buildings were built or replaced one by one, because the constructions were conducted by individual families responding to their own accommodation requirements. The old buildings were not torn down randomly. Only if they were naturally eroded to the point that they could not be simply repaired, they would be replaced by new ones. Thus, sub-settlements grew gradually.

In the third stage after 1980, the metabolism approach became a large-scale development method. A group of new buildings was built on a site once under the planning of the government. For example, the development of Xidong, as an earlier case, was initiated by the government along with the construction of the Sixiao Road. New buildings were lined along the road in uniform form and size. The development of the north part of Huayuan is a recent case of large-scale construction by the government. This project began in 2015. Nine rows of buildings were designed and built together by the government and were sold to residents who moved out of the reserved area of the cultural park. The governmental planning seems to be a well-ordered model that upgrades the settlement through prior planning from an overall perspective. But the inner logic behind the government's actions is detached from real life and the daily needs of ordinary people. The seemingly ordered model shows inner disorder in fact.

The metabolism approach of sub-settlements changes from self-construction to totally governmental planning gradually, from the inner order of life to the seeming order of the economy.

4.3.4.3 Shifting Value of Sub-settlement

The value of sub-settlement land shifts from being culture-oriented (focused on Fengshui) to being economy-oriented.

Until the 1970s, Houchi had been the best living area, featuring the highest built-up density, the longest history, and the best Fengshui, which attracted the most powerful families to compete for the right to live there. The area next to the Beijian bridge has been the major trading field with several stores selling goods to passengers for a long time. The newly constructed road network reverses the spatial structure of the village. The original peripheral building groups around Nanshan hill settled by the Chen family have become more prosperous in the village for its convenient location next to the new transport entrance from the province-level road into the village. New restaurants and hostels open by renovating old dwellings. The first and also the busiest restaurant in Xiaqiao belongs to this area. In 2014, the first boutique hotel that can be booked via the internet in the small village, operated by a couple living in Luoyang, opened here as well. In contrast to the flourishing of Nanshan, the west bank of Houchi, where two big traditional courtyard dwellings built by the elite from the rich Tang family in the Qing dynasty have been desolate since the 1990s due to its inconvenient location that cannot bring interests from tourism.

The shift of land value also leads to the transfer of the social structure. The previously powerless Chen family gained more power in the village to compete with Tang and Lin. And, stimulated by the booming tourism, all the values of spaces are transferring. The wooden pavilion bridges and traditional dwellings have lost their original functions of traffic or accommodation and have become view spots for visitors. The area next to the Beijian bridge changes into a park where new construction is not allowed, in order to protect the architectonic remains and attract tourists.

4.3.4.4 Diversification of Sub-settlement's Function

The function of sub-settlements changes from residential only to multi-functions, with different sub-settlements playing different roles in the village.

From the Qing dynasty to the planned-economy PRC, the major function of all the sub-settlements in Xiaqiao is residential, aiming to provide basic accommodation for peasants. But the seed of commercial function grows gradually. In the late Qing dynasty, some shops appear around the Beijian bridge belonging to the Qiaotou sub-settlement. In the Republican period, family workshops appear sporadically in Xidong, Qiaotou, and Beishanqiao along with the booming trade between Zhejiang province and Fujian province. In the communalized period after 1949, more functions are implanted into different sub-settlements. For example, a primary school with a public education function and a great assembly hall with a public gathering function were both built in Qiaotou; cooperative shops as part of the planned economy were built in Xidong. But all these new functions beyond residential are inserted independently, which cannot change the nature of sub-settlements as residential zones.

The actual diversification of sub-settlement functions begins in 2005. In this year, two important events happen that greatly influence the village: the construction of a new provincial road and the establishment of Langqiao Cultural Park. Improved transportation brings more opportunities for economic development. The cultural park shows the new strategy of economic growth enacted by the government. In the overall planning of Xiaqiao, the whole village is divided into different zones with different roles and functions. The whole village is differentiated by function for the first time.

4.3.5 Growing Sub-settlements in Different Contexts: Development of Houchi, Xidong, and Qiaotou

In different settings, the growth orbit of sub-settlement often shows different features. This research uses the Houchi sub-settlement as a case study to analyze different stimulations that influence the development of sub-settlement during different socio-economic periods. Besides, two other cases, namely the Xidong and Qiaotou sub-settlements, are proposed for comparative studies to further discuss the transformation approach of the built environment at this level.

4.3.5.1 Case Study: Houchi

Houchi is a typical case to explore the development process that transformed from a self-organized approach to a governmental control approach. In the Qing dynasty and the Republican China period, its development was stimulated by internal needs from the residents themselves. But in recent decades, the influence from the government has been enhanced, and the government has been greatly involved in the construction of Houchi. The current configuration of Houchi's built environment contributes to the mixture of controlling forces from both the external agent, i.e., the government, and the internal requirements of residents.

During the Qing dynasty and the Republican period, the construction in rural society kept the mode of self-organization. The real power of the government has been limited to the administrative level of the county. The large area of rural built environment was adjusted in accordance with the living requirements of the ordinary people.

Houchi was first settled by the Jinan Lin lineage. With the division of the lineage, the residents living in Houchi belong to the northern fraction of the lineage, compared to the southern fraction, which refers to the residents living in Huayuan.

Before the Qiaolong period of the Qing dynasty, Houchi was fully controlled by the Lin lineage. They built several courtyard houses there and governed a large amount of farmland next to the houses. The land of Houchi is long and narrow, lying between the Beixi River and the East Mountain. Thus, the houses are arranged along the foot of the mountain (Fig. 4.40). During that time, even the lands on the opposite bank of the river that Houchi faces directly belonged to the Lin lineage. Today, the earlier ancestral hall of the north fraction of the Lin lineage still stands there, gazing at Houchi silently, although it has been discarded and replaced by the new one built inside Houchi (Fig. 4.41).

However, at the end of the Qianlong period, the Tang lineage entered the governance of Houchi. The Tang lineage moved to Xiaqiao only in the Kangxi period of the Qing dynasty. It was the latest one of the five lineages that finally settled in Xiaqiao. According to Tang's family archives, the third generation of the lineage, Tang Kaiyou, built the courtyard house, the Old Tang House, in the southern part of Houchi (Appendix III) in the 59th year of the Qianlong period (A.D. 1794). It was the same year that the fourth son of Tang Kaiyou won the third prize in an official examination.

The success in official examinations is always a symbol that shows the prosperity of the family. It marks the great fame, power, and fortune that a family can gain in rural society. After all, in an illiterate society, knowledge manifests the power and right to join in the management of public affairs. The success in official examinations

Fig. 4.40 Houses Rowed along the Foot of the Mountain. (Based on the Drawing in the Lin's Family Archives)

Fig. 4.41 The Earlier Ancestral Hall of the Houchi Lin Lineage Division Left: Photo; Right: Location Map. (Based on the Drawing in the Lin's Family Archives)

is always a sign that the winner, as a rural intellectual, gains approval from the government. He becomes the representative from both the official and folk perspectives. For the ordinary people in rural society, he is the representative of the government to inform them of the official regulations; for the government, he is the representative of the peasants to report their interests and will.

The success in official examination has a direct relation with the construction of Tang's first big courtyard house. It is the intangible prerequisite for building a big courtyard house in rural society. It identifies the status of the Tang lineage as a significant force involved in the informal governing of rural society. It lifts the curtain on the combination and cooperation between the two lineages in governing Houchi.

Besides the intangible preparation to build a big courtyard house in Xiaqiao, accumulated wealth is the tangible prerequisite. After all, the cost to build a house with a big size, a complete courtyard form, and delicate decorations cannot be ignored. Also recorded in Tang's lineage archives, Tang Kaiyou, the organizer of the construction of the Old Tang House, married the daughter of the Zhang family living in Qianping, who was one of the wealthiest lineages at that time in Taishun County. It demonstrates that the Tang family possessed considerable wealth and fame in rural society.

The winning in the official examination, undoubtedly, brought great fame and prestige to the Tang family. Tang Wei, the winning intellectual, became the successor of the house. From the fourth generation, the Tang lineage formally entered into a period of rising. There were always family members who won official titles as rural intellectuals in each of the following three generations (fifth, sixth, and seventh generations) until the end of the Qing dynasty (Table 4.10, Appendix III). It is the mark of the prosperity of the Tang lineage. The successive official certifications are also the guarantee of the lineage's authority and fame in rural society.

In the first year of the Daoguang period (A.D. 1821), Tang Wei's elder brother of the fourth generation and his nephew of the fifth generation respectively built two big courtyard houses in Yangtou. It is the second wave of construction activities of the Tang lineage and also the beginning of lineage division for the Tang lineage. Through the dividing behavior, the Tang family formally became a lineage with five

Table 4.10 Statistics of Titled Family Members in the Tang Lineage

Generation	En Qi (恩耆)[a]		Yi Xiang Sheng (邑庠生)[b]		Guo Xue Sheng (国学生)[c]		Li Gong Sheng (例贡生)[d]		Total
	Name	Quan.	Name	Quan.	Name	Quan.	Name	Quan.	
4th	Tang Chaoyu; Tang Chaoliang	2	Tang Wei	1	/	/	/	/	3
5th	/	/	/	/	Tang Guoding; tang Guozhao	2	Tang Guoxiu	1	3
6th	Tang Fapei; Tang Fachun	2	Tang Mengxiong; Tang Mengqing	2	Tang Menglin	1	/	/	5
7th	Tang Rizhang; Tang Riguan; Tang Suding	3	Tang Gao	1	/	/	/	/	4
Total		7		4		3	‘	1	15

[a]En Qi 恩耆, Respective elderly intellectuals in the village who is titled by the county officer
[b]Yi Xiang Sheng 邑庠生, it refers to the student who study in the school of county level
[c]Guo Xue Sheng 国学生, it refers to the student who study in the school of country level
[d]Li Gong Sheng 例贡生, it refers to the student who win the title of honor by pay tribute to the government instead of attending the examination

branches named Ren, Yi, Li, Zhi, and Xin. The five male members from the fourth generation became the ancestors of the five branches.

During the whole Jiaqing period, the interval between two periods of construction activities, all people lived in the Old Tang House. There were 40 people living in the house together, including 30 adults grouped as 15 couples and 10 children (Table 4.11). One adult couple could be seen as one independent family group. Then, there were 15 families living together in the house. Considering the common living model that each family occupied one cell space, for the Old Tang House with 15 cells total, 15 families had already reached the maximum capacity of accommodation.

Too many people living together in one courtyard house might be one reason that leads to the construction of new houses and the division of lineage. The living environment is so crowded that there are insufficient living spaces. But there is another coincidence in time that Tang Wei died in the same year that the two new courtyard houses in Yangtou were built. The original family authority disappeared. It might stimulate the lineage division indirectly in a spiritual sense.

The rising of the Tang lineage reflects the decline of the Lin lineage, on the other hand. All the lands that the Tang family used to build houses originally belonged to the Houchi Lin lineage. Construction of large courtyard houses need a large amount of money. In rural society, the source of money must be farmland. From the transaction documents that the Tang family living in the Yangtou Tang House keeps, most of the land transactions relate to the Houchi Lin lineage. The Tang lineage has bought a lot of farmlands from the Lin lineage (Appendix III). Considering the irreplaceable significance that land always represents, the behavior of farmland transactions shows that the purchaser of land, the Tang lineage, has been in the wealth accumulation period, but the bargainer, the Lin lineage, has been in the vicious cycle of family finance that they can only support their livelihood by the money earned from land selling.

The conductor of building the Yangtou Tang House, Tang Guoxiu, also built the Tang's ancestral hall next to the Old Tang House in Houchi in the 26th year of the Daoguang period (A.D. 1846). Further identifying the status of the Tang lineage in the village aims to compete with the Lin lineage for the residential right in Houchi. The construction of the ancestral hall is a great move for the whole Tang lineage as a complete and formal lineage system. Besides, Tang Guoxiu has actively involved himself in public affairs in the village. He is one of the major conductors in rebuilding the Beijian bridge. It demonstrates the significant status of the Tang lineage in the village.

Table 4.11 Statistics of Resident Quantity Living in the Old Tang House until A.D. 1821

Generation	Adult		Couple	Children		Total
	Male	Femal		Male	Female	
4th	3	3	3	/	/	6
5th	11	11	11	/	/	22
6th	1	1	1	6	4	12
Total	15	15	15	6	4	40

The continuous construction of houses in Houchi has not ceased. In the seventh year of the Tongzhi period (A.D. 1868), the biggest house of the Tang's family, the Huoqiangdi House, was built in the middle part of Houchi by Tang Zhao, the little cousin of Tang Guoxiu, who was also an intellectual from the fifth generation. This site originally belonged to the Lin lineage as well. Until today, three independent two-story houses and one one-story house have been built right behind the Huoqiangdi House.

But the Huoqiangdi House was not built all at once. In the Tongzhi period, in the first construction, only the main building with five cells was built. The left and right wing buildings were built by Tang Gao, an intellectual who won the official examination from the seventh generation, in the 25th year (A.D. 1899) and 33rd year (A.D. 1907) of the Guangxu period, respectively. The stone walls with round courtyard gates were constructed after that. The construction of the Huoqiangdi House is the peak of the development of the Tang lineage. Now, it is still the largest dwelling in Xiaqiao.

In contrast to the booming of the Tang lineage, the Houchi Lin lineage has not built big courtyard houses. By comparing the maps remained in the two family archives, written in the 14th year of the Jiaqing period (A.D. 1804) and the 12th year of the Republican China period (A.D. 1923), there have been built three independent two-story houses and one one-story house. It is in sharp contrast with the ceaseless construction of several big courtyard houses by the Tang lineage during the same period.

Instead of Houchi, the Lin lineage even lost the dominance right over the Yangtou sub-settlement, which lies on the other bank of the Beixi River and faces Houchi. The ancestral hall of the Lin lineage, built in the Qianlong period, is sandwiched between two big Tang's courtyard houses. The loss of control over the Yangtou area might be one reason for the current abandonment of the Lin ancestral hall.

Although the Tang lineage and the Houchi Lin lineage competed for the living resources in Houchi, they also cooperated to govern Houchi together in another sense. Taking the Xiyuan Lin divisions, which belong to the same northern branch of the Jinan Lin lineage but are different divisions, as an example, we can compare their marriage relations with the Tang lineage to those of the Houchi Lin division (Table 4.12, Appendix III). From the Tang's family archives, although the number of marriages is fewer than average, there are still some marriage relations between the two lineages. It reflects that in the competition for governing Houchi, the two lineages still maintained a positive relationship with individual interactions, instead of fighting and being completely isolated.

At the end of the Qing dynasty, the pattern that the Lin lineage controlled the northern part of Houchi and the Tang lineage controlled the southern part finally formed. This pattern has been mostly inherited by the following periods (Fig. 4.42).

In the Republican period, the land became fragmented, changing from several large built sites to more small plots. There are few buildings of smaller size in Houchi due to the unstable social environment. The new houses were all built in the spare spaces surrounding the original ones. Only two houses were built in the 34-year period by the Tang lineage in Houchi. Both were built around the Old Tang

Table 4.12 Marriage relation with the Tang Lineage

Generation	Houchi Lin			Xiyuan Lin		
	Male	Female	Total	Male	Female	Total
3rd	1	/	1	/	2	2
4th	/	/	/	/	/	/
5th	2	/	2	/	/	/
6th	1	/	1	/	2	2
7th	/	2	2	1	2	3
8th	1	/	1	2	2	4
9th	1	/	1	1	1	2
Total	6	2	8	4	9	13

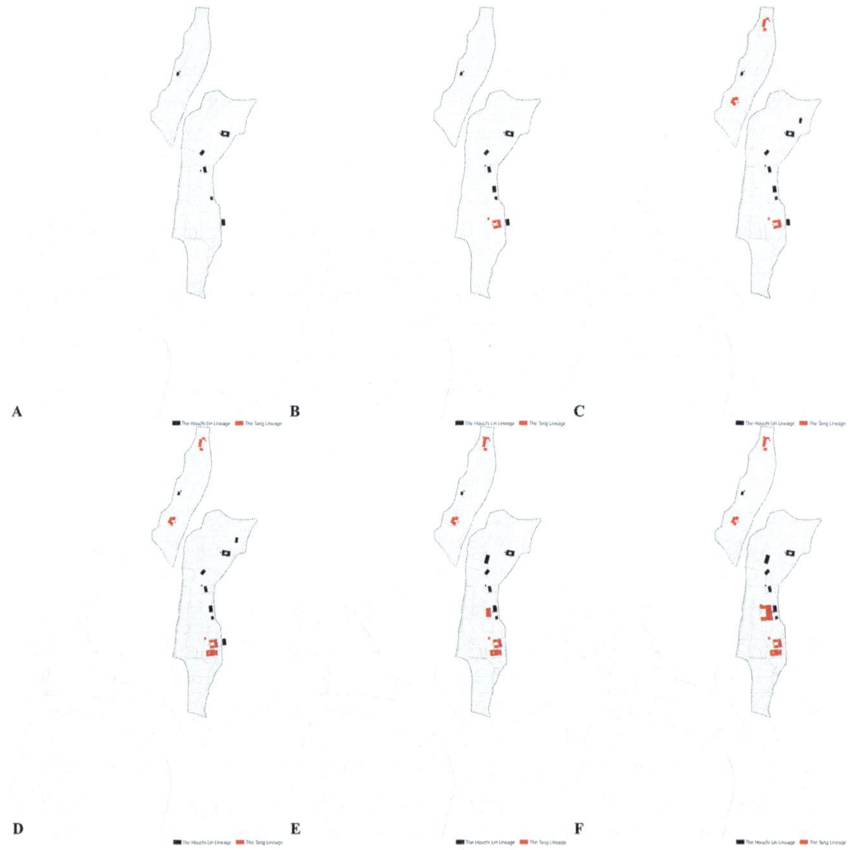

Fig. 4.42 Development sequence of Houchi and Yangtou in the Qing Dynasty: (**a**) 1644–1793; (**b**) 1794–1809; (**c**) 1810–1821; (**d**) 1822–1846; (**e**) 1847–1868; (**f**) 1869–1911

House by residents who moved from it. One is a three-story building connecting to the Old Tang House in the backyard, and the other is an independent house built with stone walls to the north of it. The three-story house is the highest building in Houchi at that time. It still reflects the status of the Tang family.

The general spatial pattern of Houchi has been maintained, with the Lin lineage living in the northern part and the Tang lineage living in the southern part. However, compared with the Qing dynasty, the middle part of Houchi in the Republican period became more mixed and complicated, where the Lin lineage and the Tang lineage lived together.

Due to the weaker direct management of the village by the government, the growth approach of Houchi still belonged to a self-organized approach responding to the requirements of residents' daily life. At the end of the Republican period, with the actual land occupation by the Lin and Tang lineages, the spatial pattern finally formed, with the northern part occupied by the Lin lineage, the southern part occupied by the Tang lineage, and the middle part sandwiched between them co-occupied by both lineages together (Fig. 4.43).

The official force from the government gradually permeated into the spatial organization of the sub-settlement by controlling different steps in the building process. In the planned-economy period (1949–1979), the government controlled rural housing through supervision to decide the distribution and redistribution of farmland, homesteads, and existing houses. However, in the market-economy period (after 1980), the intangible hand of the government has gradually reached every corner of rural space and every step of the building process to launch and organize real construction projects.

The direct management from the government enhanced greatly after the establishment of the PRC due to the complete change in the political and administrative system. Administrative efficiency has improved. The official governance has not stopped at the county level but has a direct impact on the village itself, and the governmental

Fig. 4.43 Development sequence of Houchi and Yangtou in the Republican Period: (**a**) 1912; (**b**) 1913–1929; (**c**) 1930–1949

power has been strengthened over time. Every village has a village committee as a semi-governmental organization to assist in dealing with public affairs in rural settlements.

Instead of individual transactions of land among the villagers themselves, all the newly added homesteads that might be converted from other lands, such as farmland, forests, and mountains, need to be officially approved by the government. But after the approval of the homesteads, the construction process was organized by the residents themselves beyond the control field of the government.

For the existing houses, the government was involved in the redistribution process as a strategy to fulfill the ideal of equalizing wealth. In the land reform, the owners of the Huoqiangdi House were treated as landlords. The big houses were redistributed to other villagers who did not originally live in the houses. It was a direct intervention by the government in housing under the unique ideology.

Besides, Houchi, as a unit of the production brigade, created public spaces shared by all members in the brigade. For example, a public warehouse was built in the 1960s to store production resources together. Few private houses were built during this period. The ideology of equalization to share resources in a community is clearly reflected in the transformation of the spatial pattern of the sub-settlement (Fig. 4.44).

After the 1980s, the built environment of Houchi changed greatly (Fig. 4.53). Instead of increasing smoothly, Houchi experienced a sudden construction boom between 2005 and 2007. This referred to the row-like houses neatly lined along the main road in Houchi. It was the result of a governmental adjustment to settle residents who lost their houses in the typhoon and flood of 2005. The scale of construction area in just two years had never been seen in the small village before. Only the powerful force from the government could realize it.

Besides a larger scale of construction, the spatial pattern of the newly constructed houses differs from that of previous ones. Instead of courtyard houses or independent houses, the new houses are row-like and neatly lined along the road (Fig. 4.45). They

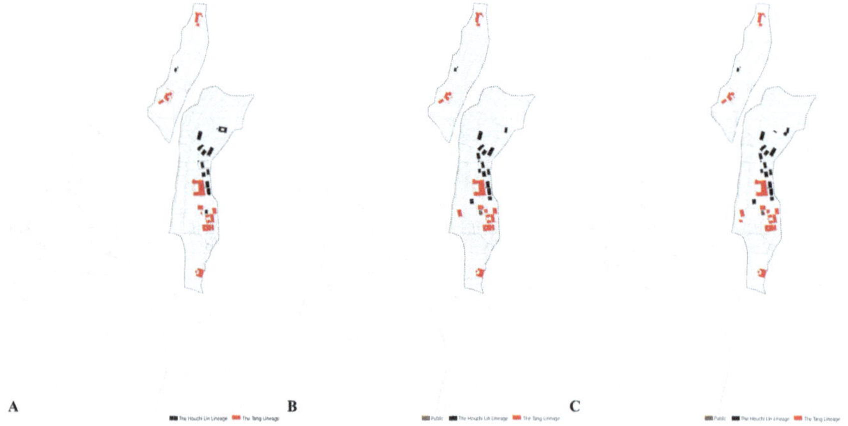

Fig. 4.44 Development sequence of Houchi and Yangtou in the Planned-economy Period: (**a**) 1949; (**b**) 1950–1969; (**c**) 1970–1979

are relocated next to new lines of transport, which substitute the original group-shaped building fabric. It is a new spatial configuration that shows rural housing here is emancipated from the restriction of farmland preservation. The row-like pattern fits the requirements of the increasing mobility of rural populations as well as their accessibility to more resources. Meanwhile, transitional space has disappeared as well. The doors of new houses directly face the road and facilitate daily access. It fosters the spatial transformation from living spaces to small groceries next to the road where goods are sold to passengers and drivers.

House orientation changes as well. Before the 1980s, the orientations of all houses in Houchi were unified, facing westward toward the river with the mountains lying behind. It is based on the traditional method to fit the geographical conditions, which is a narrow space sandwiched between the river and the mountain. The adaptation of terrain then evolved into a complete geometric theory as Fengshui. This theory strengthens the lucky impression of the land to live on. The locals believed that residing in Houchi could bring prosperity to the specific family who lived there. But orientation changes occurred in the new buildings. The new buildings have a totally opposite direction, facing east toward the mountains and with their backs to the river on the west side. The changes result from the construction of the road. The new buildings are located on the opposite side of the road, next to the river instead of the mountains. The orientation depends on the road instead of the geographical condition, so the new buildings, indeed, face the road.

Along with the change of space, the programming of houses in Houchi has been diversified. In the past, Houchi was mainly for residential purposes. But after the 1990s, multiple functions were mixed together. More commercial functions emerged here. Small groceries opened along the road. In 2016, a new hotel named 'Lao Cuo Yi Hao' (老厝一号) opened as a result of the booming tourism in the whole Xiaqiao Village.

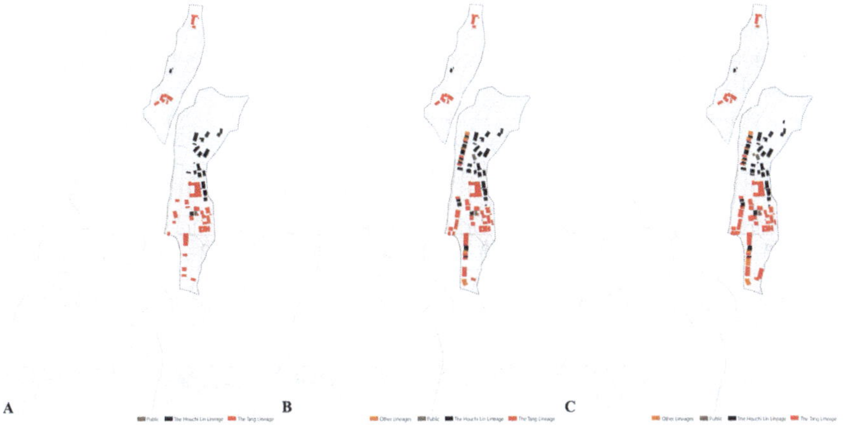

Fig. 4.45 Development sequence of Houchi and Yangtou in the Market-economy Period: (**a**) 1980–1999; (**b**) 2000–2009; (**c**) 2010–2016

Besides the tangible changes in space, the residential constitution has been changed as well. The original pattern of residential zones, dominated by lineages, has been changed. On the one hand, in the newly constructed places, the Tang lineage and the Lin lineage are living together. On the other hand, other lineages are allowed to live here under the regulation of the government. Most of them live in the peripheral area of Xiaqiao. They are encouraged to move out from their original residences due to the policies of village combination.

All the tangible and intangible changes in the housing of Houchi reflect the transitioning control mode of space in the village. The approach of governmental adjustment gradually replaces the original self-organized way of constructing in rural society throughout the whole building process (Fig. 4.46).

After the new round of land reform in the 1980s, the control of lands and constructions by the government has been strengthened again. The approval of newly added homesteads has become more strict. Villagers who want to build new houses need to submit an application to the town government. The application will be handed to the higher-level county government and will wait for the quota of new construction areas under the consideration of the whole county. It is impossible for them to acquire a new homestead to build new houses by themselves under the strict regulation of land control after the 1980s land reform.

The government has absolute power to control the usage of all lands. Instead of ownership, common peasants only have the right to use the lands in a set way delimited by the government. The situation deteriorates for the landless farmers whose lands have been expropriated by the government for other purposes, mostly for the construction of infrastructure.

The usage of land in villages has been planned by the government on a larger scale. Instead of only responding to the needs of the locals, the government considers more about regional development with other areas. For example, since 2016, a new road has been constructed to the north of Houchi. It is a new provincial road passing by Xiaqiao to Taishun County and then directly to neighboring Yunhe County. Under the planned road network, the lands in Houchi need to be expropriated from the locals by the government. The provincial road obviously has no direct relation to a common peasant who lives in Houchi and farms in the lands within a distance of 500 m. But it

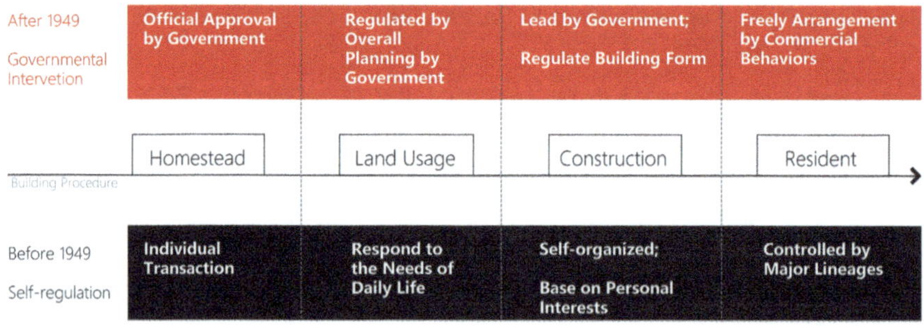

Fig. 4.46 Two modes of construction along the building procedure

responds to the regional development under the scope of the whole Taishun County. Thus, through governmental intervention, the space inside the village acquires other meanings beyond the village.

In real construction projects, the involvement of the government has been more obvious since 1980, sometimes dominating housing projects. The government plays an active role in launching and organizing construction projects, rather than mere passive supervision. During the planned-economy period, the government's approach to regulating housing projects was simple, which was by approving the applications submitted by the villagers. After obtaining legal proof of land, houses were still built by the villagers themselves. But after 1980, the government could decide on the new housing area and start the project by itself. For example, the new housing area along the road in Houchi was proposed by the government on its own initiative, and then opened to individual applications.

Moreover, after 2010, the government even involved itself in the real construction process of building projects. For example, the Daligang residential area was constructed by the construction team hired by the government officially. It completely stripped the villagers of the self-housing model that had lasted for hundreds of years.

The appearance of buildings is another aspect that the government has started to control. Since Sixi Town was identified as a historical town in the fourth evaluation in 2012, there are official regulations in different planning documents, such as protection planning and detailed planning, that clearly regulate the height and form of buildings. Villagers cannot build houses freely according to their own preferences. Besides, along with the booming tourism, the government pays more attention to the appearance of existing buildings. Since 2015, the houses along the main road in Houchi have been decorated uniformly, adding traditional patterns and decorations on the façade facing the street. The control of architectural form is a new field that the government has started to engage in.

In the government-dominated mode of construction, the development of sub-settlements has not been closely related to the locals' daily needs. Instead, it directly reflects the requirements of governmental behavior. Meanwhile, the housing problem has changed from a problem of basic living resources for ordinary people to commercial behaviors guided by the real estate market.

4.3.5.2 Comparative Study: Xidong and Qiaotou

Xidong and Qiaotou are two sub-settlements where the built environment has transformed greatly from the Qing dynasty to the present under the influence of changing development modes. However, the different settings these two cases face have brought about completely reverse processes of built environment transformation (Table 4.13).

Both Xidong and Qiaotou are traffic hubs next to a river with a pavilion bridge inside. Xidong is at the eastern bank of the Dongxi bridge crossing the Dongxi river on the southern boundary of Xiaqiao, which neighbors the Baifenqiang village. Qiaotou

Table 4.13 Comparison of Xidong and Qiaotou in the Qing Dynasty, Republican Period, the Planned-economy Period and the Market-economy Period

Sub-settlement		Xidong	Qiaotou
Geographical location	Location	Southern part of Xiaqiao	Central part of Xiaqiao
	River	Next to the Dongxi river	Next to the Beixi river
	Bridge	Xidong bridge	Beijian bridge
	Terrain	Narrow highland along the river	Large plain area on the bank
Occupied lineage	Mode	Single-lineage occupation	Multi-lineages occupation
	Specific lineages	The Chen lineage	The Chen lineage (major); the Xihejun Lin lineage (major); The tang lineage
Role		Peripheral for Xiaqiao; but important for Baifenqiang	Center of Xiaqiao; important for Xaqiao
Function	1st	Traffic hub	Traffic hub
	2nd	Religious	Trade center
Main building		Temples; houses	Shops; houses
Development	First developed	Later	Earlier
	Initial buildings	Religious architecture (the Linshui Temple; the Chen Daweng Temple)	Houses
	First developed period of housing	Housing began later only from the Guangxu period, the end of the Qing dynasty	Housing began from the Kangxi period, the early of the Qing dynsty
Sub-settlement		Xidong	Qiaotou
Planned economic PRC (1949–1979)	Role	Peripheral area in Xiaqiao	Central area in Xiaqiao
	Function	Residential; commercial	Residential; commercial; administrative; education
	Type of new buildings	Cooperative of Xiaqiao production brigade; houses	Grand hall; primary school; Office of Xiaqiao Production Brigade; houses
	Occupied lineage	The Chen lineage	The Chen lineage (major); the Xihejun Lin lineage (major); The Tang lineage

(continued)

Table 4.13 (continued)

Sub-settlement		Xidong	Qiaotou
	Major approach of land distribution	Land redistribution by government	In accordance with original homesteads
	Ownership	Public: Cooperative; temples (using as brigade kitchen) Private: Houses	Public: Grand hall; primary school; administrative office Private: Houses; shops
1st period of market-economy period (1980–2004)	Role	Common residential area	Earlier
	Function	Residential; commercial; religious	Residential; commercial; administrative
	Type of new buildings	Private shops; houses	Grand hall; Office of Village Committee; houses
	Occupied lineage	The Chen lineage (major); the Tang lineage; the Xihejun Lin lineage	The Chen lineage (major); the Xihejun Lin lineage (major); The Tang lineage
	Major approach of land distribution	Transact homestead from government	In accordance with original homesteads
	Major approach of house construction	Self-construction in accordance with governmental regulation	Self-construction in accordance with governmental regulation
	Ownership	Public: Temples (using as brigade kitchen) Private: Houses; shops	Public: Grand hall; administrative office Private: Houses; shops
2nd period of market-economy period (2005–2016)	Role	Major commercial street	Central area in Langqiao Cultural Park
	Function	Residential; commercial	Toursim; residential; commercial; administrative
	Type of new buildings	Private shops; houses	/
	Occupied lineage	The Chen lineage; the Tang lineage; the Xihejun Lin lineage; the Jinanjun Lin lineage; several families from other places	The Chen lineage (major); the Xihejun Lin lineage (major); The Tang lineage

(continued)

Table 4.13 (continued)

Sub-settlement		Xidong	Qiaotou
	Major approach of house construction	Strictly regulated by government; unified decoration of street façade by government	Forbidden to build new buildings by government
	Ownership	Public: Temples (using as brigade kitchen) Private: Houses; shops	Public: Grand hall; administrative office Private: Houses; shops

is at the center of Xiaqiao, that covers both banks of the Beijian bridge crossing the Beixi river.

The similarities in geographical location bring similar initial conditions to their development. They all serve as traffic hubs on the must-pass routes running through the adjacent villages. Xidong connects Xiaqiao with Baifenqiang, serving the people from the south or heading to the south, and Qiaotou connects two banks inside Xiaqiao, serving the people from the north or heading to the north. Due to the core role of the bridges inside the two sub-settlements respectively, all the buildings were built around the bridge at first. The bridge is the starting point of their development. For Xidong, two temples, the Linshui temple and the Chen Daweng temple, were built next to the Xidong bridge; for Qiaotou, the houses and shops were built surrounding the Beijian bridge on both banks.

Moreover, the differences in geographical location decide the different development sequences that follow. It is located in the peripheral area of Xiaqiao and has not been well constructed by Xiaqiao's residents throughout the Qing dynasty. However, for Baifenqiang, Xidong occupies a very important location due to the flow direction of water. It sits on the water end of Baifenqiang. Water always symbolizes fortune, thus the rural residents believe that controlling the end of the river means preserving the fortune of the village from loss. People always built bridges and temples to guard the water end. According to the inscribed stone tablet (Fig. 4.47), the sponsors who financed these three religious constructions were mostly from Baifenqiang, though some were from other adjacent villages. Meanwhile, for Qiaotou, it is at the center of Xiaqiao that connects different parts of Xiaqiao, directly benefiting the daily life of the locals. Thus, a lot of houses were built in this area gradually. Due to its significant location, it became a trading area in the village with the earliest shops opening on its two banks. The shops even connect to the bridge directly as a whole (Fig. 4.48).

Xidong and Qiaotou faced different situations of lineage occupation during the Qing dynasty. Xidong was a sub-settlement dominated by a single lineage, the Chen family. During the Guangxu period, a Chen family who originally lived in Nanshan moved here and built their new house next to the Chendaweng Temple. A dyehouse was built behind the house later.

Comparatively, Qiaotou is a mixed residential area occupied by multiple lineages, including the Xihejun Lin lineage, the Tang lineage, and the Chen lineage. They all

Fig. 4.47 Inscribed stone tablet of the Dongxi bridge

Fig. 4.48 Spatial relation between bridge and accessorial shops

built houses in this sub-settlement. The co-settled mode is due to the significance of Qiaotou in the village. One of the most important constructions, the Beijian Bridge, was built in this area. It is not only the landmark of the village, but also a key traffic hub that must be passed by to go across the Beixi River. The construction, reconstruction, repair, and daily maintenance are financed by all five lineages in the village together (Fig. 4.49). Thus, all the lineages have the right to own a piece of land surrounding the bridge. Today, we can see buildings belonging to different lineages standing in Qiaotou. Among them, the Chen lineage and the Xihejun Lin lineage are the two major ones. The Chen lineage, whose ancestor Chen Ruchang was one of the main conductors organizing the building activity of the bridge, occupied the area on the east bank of the bridge. The Xihejun Lin lineage occupied the west bank of the bridge which operated shops to earn a living initially.

Xidong and Qiaotou followed different development sequences (Fig. 4.50, Fig. 4.51). Xidong's development lagged behind the other sub-settlements iniao. When the other sub-settlements had been settled by families from the five lineages and dozens of houses had been built, only two temples were built next to the Xidong bridge in the southern corner of Xidong. Then for a long time, there were seldom constructions in Xidong until the end of the Qing dynasty, when a house of one Chen family was built next to the temples.

This is related to the topography of Xidong. Most of the land there was occupied by a small mountain that is not suitable for construction. Besides, only a narrow space along the Dongxi River has been left that contains a muddy path connecting Xiaqiao and Baifenqiang, and a large amount of farmlands.

Comparatively, Qiaotou was inhabited by villagers as a major residential area in the earlier period of the Qing dynasty. At present, the earliest house built in Qiaotou that has been recorded is the house of Chen Ruchang, one of the four initiators to

Fig. 4.49 Inscribed stone tablet of the Beijian bridge

Fig. 4.50 Development sequence of Qiaotou and Xidong in the Qing Dynasty: (**a**) 1644–1820; (**b**) 1821–1874; (**c**) 1875–1911

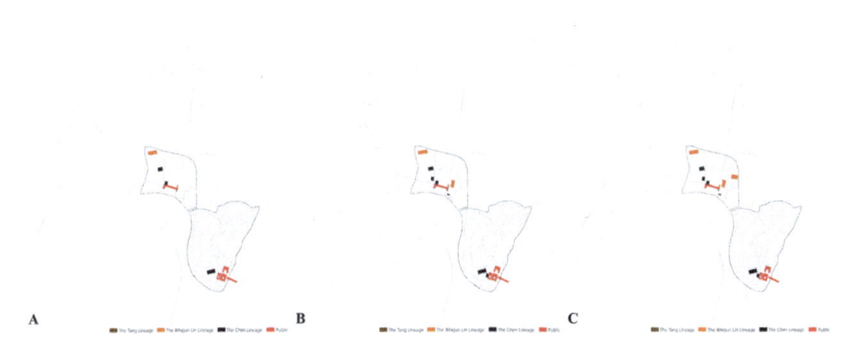

Fig. 4.51 Development sequence of Qiaotou and Xidong in the Republic Period: (**a**) 1911–1919; (**b**) 1920–1939; (**c**) 1940–1949

organize the construction of the Beijian bridge, on the western bank, built in the Kangxi period. After the construction of the bridge, the area of Qiaotou became a busy "traffic station" where passengers flocked to have a rest. Due to the daily flow of passengers passing here, it provided a great opportunity for people to trade goods here. Thus, Qiaotou can also be seen as an embryo of a commercial area in rural society during pre-modern times. More buildings were built surrounding the bridge in Qiaotou. Most of them were for both residential and commercial purposes.

After 1949, following the establishment of the PRC, the development sequences of Xidong and Xiaqiao were also different, although the government has had a considerable impact on both of them (Figs. 4.52 and 4.53). Their development is closely related to governmental planning. Based on the degree of government involvement, 2005 is a turning point when the Langqiao Cultural Park was established. After that, the two sub-settlements transformed rapidly.

Xidong's development is obviously promoted by the government. Before the 1970s, Xidong was still not been a major residential quarter in Xiaqiao, where few houses had been built. During the collectivization period, a cooperative shop was built in the middle of Xidong, selling sundry goods for Xiaqiao's villagers. In the 1970s, a modern road was built. It passed through Xidong and connected the north-eastern area with the southern area in the county. It was the warm-up period of an overall development of Xidong. The lands on both sides along the road were redistributed to the qualified candidates who applied for a plot of land for housing during this period. But large-scale house construction had not started.

The formal launch of Xidong's development began in the 1990s. The road passing through Xidong has been reconstructed and upgraded. It directly connects to Baifenqiang, where the town government is located. Then the government decided to build its new offices between Baifenqiang and Xiaqiao. They finally chose the place next to Xidong along the renovated road, Sishi West Road. It would be formed as the new center of the town. Then, a large number of houses started to be built along the road around the new town center. This force greatly influenced the development of Xidong, so lined houses were built along the road on both sides. Xidong became a new residential quarter in Xiaqiao rapidly.

Besides, during the 1990s, the Second Middle School of Taishun was built on the highland of Xidong. This school is an important educational resource in Taishun County, serving the whole southern part. In order to facilitate study, residents from other villages rent houses in Xiaqiao.

The current configuration of Xidong was formed in 2005 when the Langqiao Cultural Park was officially established. According to the planning of the cultural park, Xidong has been identified as the major commercial street for tourism. More

Fig. 4.52 Development sequence of Qiaotou and Xidong in the Planned-economy Period: (**a**) 1949–1959; (**b**) 1960–1969; (**c**) 1970–1979

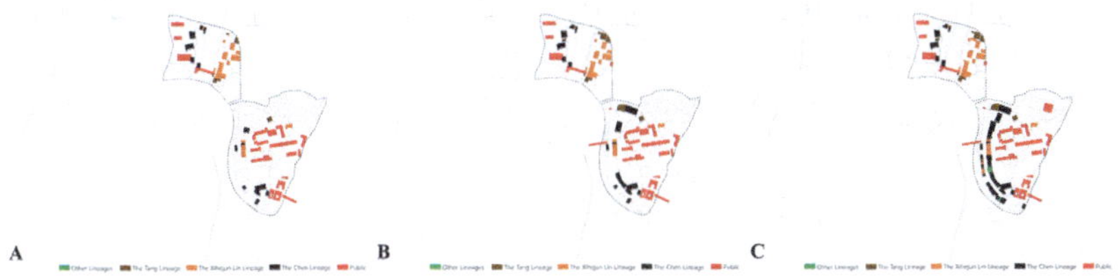

Fig. 4.53 Development sequence of Qiaotou and Xidong in the Market-economy Period: (**a**) 1980–1999; (**b**) 2000–2009; (**c**) 2010–2016

buildings have been built, filling in the previously blank areas. Many residents use the first floor to sell souvenirs and local food. A commercial street is formed under the adjustment and planning of the government.

There are strict regulations to control the architectural appearance, including height, form, etc. All the facades facing the street are decorated by the government, imitating traditional dwellings. From 2015, a new renovation project organized by the government, which aimed to transform the roofs of the houses along the street from flat to traditional sloping roofs, was conducted.

Thus, for the Xidong case, the government is the stimulus that fosters a series of developments in different stages. However, for the Qiaotou case, the government can be both a catalyzer and an inhibitor for sub-settlement development.

Qiaotou has been the center of the village for a long time. In the planned-economy period, a grand hall was built next to the Beijian bridge by the production brigade. It created a public gathering space for the whole village. It is also the new landmark that reflects the central role of Qiaotou in the village. The supersedure that happens in Qiaotou, replacing the old buildings with new ones, follows the self-organized order that depends on residents' original homesteads. Only those who construct new buildings need to apply to the government and wait for approval of new homesteads.

The turning point of the approach that the government engages in the development of Qiaotou is also from the official establishment of the Langqiao Cultural Park in 2005. In contrast to the case of Xidong, the development of Qiaotou has been limited according to the official regulations released by the government.

In the planning, the whole Qiaotou area is covered inside the park, which belongs to a protective area of the Beijian bridge. All kinds of new construction activities are strictly restricted, even for private housing. In order to facilitate management, the government planned to move out all the residents inside the area and finally transform Qiaotou into a complete tourism area. But the local residents have not agreed on the way that the government planned. They do not want to move out from their original houses where their families have lived for hundreds of years. Still, new buildings are not allowed to be built there. Thus, there are some areas discarded in Qiaotou. They are the sites of the previous houses destroyed by fire. Due to the control of the government, villagers cannot build new ones on the original sites.

In general, both Xidong and Qiaotou are the areas whose development has been greatly influenced by the government after 1949, especially in the recent decade. But, according to the different settings they face, the control of the government plays different roles specifically.

4.3.6 Summary of the Third Level

The sub-settlement level is the third level of analysis, which describes the spatial pattern of the different zones inside Xiaqiao village. The village is composed of seven sub-settlements. The evolution of rural settlements is characterized by a multi-faceted nature, given the varied rates and extents at which different sub-settlements progress concurrently. The metabolic process comprises three distinct stages: self-construction, government intervention, and overall planning by the government. The land value undergoes a transition from a culture-oriented (with Fengshui being a key cultural factor) perspective to an economy-oriented perspective, accompanied by a shift in its function from residential to multi-functional. This transformation is influenced by the distinct roles played by the various sub-settlements within the village.

At the rural sub-settlement level, the development of different areas is more complex than ever, although the social relations are much simpler and the occupied area is much smaller than in urban communities. A discrepancy has been observed in the development of the built environment among different sub-settlements. It has been demonstrated that they progress at a distinct velocity and to a varying degree, concurrently. Within the same sub-settlement, the development rate is not constant. During the internal development period in premodern China, including the Qing dynasty, the Republican period, and most of the planned-economy period after the founding of the People's Republic of China (PRC), the differences in development among various sub-settlements were not apparent. A significant increase in polarization became evident following 1980, a period that witnessed an escalation in government intervention and control. The construction of certain areas was executed at an unprecedented rate. This large-scale construction rapidly disrupted the prevailing settlement patterns. In addition to the rapid increase in built-up area, regulations that previously controlled traditional folk construction were relaxed in order to unify buildings under official planning. The metabolic approach to settlements underwent a gradual transition from self-construction to total government planning, reflecting a shift from the inner order of life to the apparent order of the economy.

Refereneces

1. Vitor Oliveria. (2016). Urban Morphology: An Introduction to the Study of the Physical Form of Cities, Switzerland: Springer, pp. 11.
2. Editorial Committee of the Encyclopedia of China中国大百科全书总编辑委员会. (2016). Zhongguo Dabaike Quanshu: Zhongguo Dili 中国大百科全书:中国地理. (Encyclopedia of China: Chinese Geography). Beijing: Zhongguo Dabaike Quanshu Chubanshe.
3. Skinner G. William, Mark Henderson. (2007). Physiographic Macroregions of China Proper.
4. Wu, Songdi 吴松弟, Liu Jie 刘杰. (2009). Zouru Zhongguo de Chuantong Nongcun 走入中国的传统农村:浙江泰顺历史文化的国际考察与研究(An International Exploration and Research on the Historical Culture of Taishun, Zhejiang: Stepping into Traditional Rural China). Jinan: Qilu Publish. pp. 38–39.
5. Tu Dingnai 涂鼎鼐, Bao Dafang 包大方. (1633). Ming Chongzhen Liunian Taishun Xianzhi 明崇祯六年泰顺县志 (Taishun Xianzhi of the 6th year of Chongzhen Period, the Ming dynasty), Collected by Wenzhou City Library.
6. (1685) Wenzhou Fuzhi (Kangxi Ershisi Nian) 温州府志(康熙二十四年) (Wenzhou Fuzhi of the 24th year of Kangxi Period, the Qing dynasty), Collected by Wenzhou City Library.
7. Zhu Guoyuan 朱国修, Zhu Tingqi 朱廷琦. (1729). Xuxiu Taishun Xianzhi (Qing Yongzheng Qinian) 续修泰顺县志(清雍正七年) Xuxiu Taishun Xianzhi of the 7th year of Yongzheng period, the Qing dynasty, Collected by Wenzhou City Library.
8. Lin E 林鹗, Lin Yonglin 林用霖. (1865). Taishun Fenjianglu of 分疆录 (Field Division of Taishun of the 4thyear of the Tongzhi period, the Qing dynasty), Collected by Wenzhou City Library.
9. Zong Yuanhan 宗源瀚. (1893). Zhejiang Quansheng Yutu Bing Shuilu Daoli Ji 浙江全省舆图并水陆道里记 (Land and River Map of the Zhejiang Province), the 19th year of the Guangxu period, Collected by Wenzhou City Library.
10. (1991). Wenzhoushi Jiaotong Youdianzhi Congshu: Taishunxian Jiaotongzhi 温州市交通邮电志丛书:泰顺县交通志(Transportation and Mail of the Wenzhou City: Transportation Archives of the Taishun County). pp.11.
11. Zong Yuanhan et al. 宗源瀚等. (1893). Zhejiang Quansheng Yutu Bing Shuilu Daoli Ji浙江全省舆图并水陆道里记 (A Comprehensive Map of Zhejiang Province with Records of Water and Land Routes and Distances).
12. Editorial Committee for County Annals of the Taishun County《泰顺县志》编纂委员会. (1998). Taishun Xianzhi泰顺县志 (County Annals of the Taishun County).Hangzhou: Zhejiang People's Publishing House.
13. Pu Xincheng 浦欣成. (2012). Chuantong Xiangcun Juluo Erwei Pingmian Zhengti Xingtai de Lianghua Fangfa Yanjiu传统乡村聚落二维平面整体形态的量化方法研究 (Quantitative Research on the Integrated Form of the Twodimensional Plan to Traditional Rural Settlement). Hangzhou: Zhejiang University. pp.53
14. Huang Weijun 黄为隽, Shang Kuo 尚廓, Nan Shunxun 南舜熏. (1992). Minyue Minzhai 闽粤民宅 (Dwellings in Fujian and Guangdong Provinces). Tianjin: Tianjin Kexue Jishu Chubanshe. pp.116
15. Lu Qi 陆琦. (2000). Guangdong Minju 广东民居 (Dwellings in Guangdong Province). Beijing: Zhongguo Jianzhu Gongye Chubanshe. pp. 38.
16. He's Family Archives in the Taoxi Village 桃溪村何氏族谱, Vol. 7, Figure of Rural Dwellings 乡宅图
17. Freedman, Maurice. (1966). Chinese Lineage and Society: Fukien and Kwangtung. London: Athlone Press.
18. Freedman, Maurice. (1965). Lineage organization in southeastern China. No. 18. London: University of London, Athlone Press, 1965.Maurice Freedman, Lineage Organization in Southeastern China.
19. Wang Wei 王伟, Wang Jianguo 王建国 and Pan Yongxun 潘永询. (2016). Kongjian Yinhan de Zhixu——Tulou Juluo Xingtai yu Quyu he Minxi de Guanlianxing Yanjiu空间隐含的秩序——土楼聚落形态与区域和民系的关联性研究 (The Order Implied in Space: A Study on the

Correlation between Tulou Settlement Morphology, Regions, and Ethnic Groups). Architect 01: 95–103.
20. Gao, Zhenming 高珍明, Wang Naixiang 王乃香, and Chen Yu 陈瑜. (1987). Fujian Minju福建民居 (Dwellings of the Fujian Province), Beijing: Zhongguo Jianzhu Gongye Chubanshe.
21. Huangzuo黄佐. (1557).Guangdong Tongzhi广东通志(Guangdong Provincial Gazetteer).
22. Fei, Xiaotong费孝通. (1985). Xiangtu Zhongguo 乡土中国 (Rural China); Beijing: Shenghuo Dushu Xinzhi Sanlian Shudian.
23. Zhejiangsheng Taishunxian Diming Weiyuanhui 浙江省泰顺县地名委员会. (1986). Zhejiangsheng Taishunxian Dimingzhi 浙江省泰顺县地名志 (Gazetteer of the Taishun County in Zhejiang Province), Collected by the Zhjiang Province Library, pp.291-pp.297

Chapter 5
The Governance of Settlement Space (Volume I)

This chapter examines the governance mechanisms that control the transformation of the rural built environment at the macro-spatial levels. It focuses on how different agents regulate land use, public space, and settlement morphology. The analysis begins by defining the concept of governance in the rural Chinese context. It then investigates the specific governance dynamics at the Regional Level, the Settlement Level, and the Sub-settlement Level. Following this, the chapter summarizes the historical shift of macro-control power from intermediate organizations to the state government. Finally, it discusses the limitations of macro-control in regulating micro-level spatial practices.

5.1 The Concept of Governance in Rural China

The village functions as a terminus within the extensive administrative system of China. In comparison to urban centers, the village constitutes a rudimentary social structure, characterized by a relatively simple organizational framework. Indeed, it is dynamic and subject to constant change. This phenomenon represents a pivotal confrontation between tradition and modernity.

The built environment is comprised of two components: physical forms, including buildings, streets, and infrastructure, and the people who interact with these elements. In the context of the built environment, an "agent" is defined as any individual or group that interacts with and modifies physical forms, such as buildings and infrastructure. Agents exercise control over the environment by altering it to suit their preferences or maintaining its condition. This concept encompasses both the creators who construct the forms and the users who inhabit them, as both roles involve active intervention that leads to spatial transformation.

X. Lin and H. Wang, *A Multi-Level Analysis of China's Rural Built Environment*, Urban Sustainability, https://doi.org/10.1007/978-981-95-4945-0_5

The fundamental principle underlying this approach to environmental perception is the concept of control, which refers to the capacity to modify specific components of the environment. Agent groups, operating discreetly behind the scenes, have the capacity to effect profound transformations in the built environment. This phenomenon is inherent to the processes of transformation that are occurring within the rural built environment.

Three groups of agents are engaged in modeling the built environment. These entities include the government, the individual (family or person), and intermediate organizations (Figs. 5.1 and 5.2). The state and the individual constitute the two fundamental poles of society. The state, in its capacity as the formal system of governance, embodies the prevailing ideology and exerts control over society. In contrast, the individual represents the daily life of real people. Intermediate organizations serve as transitional entities, facilitating the connection between the local government and individual villagers.

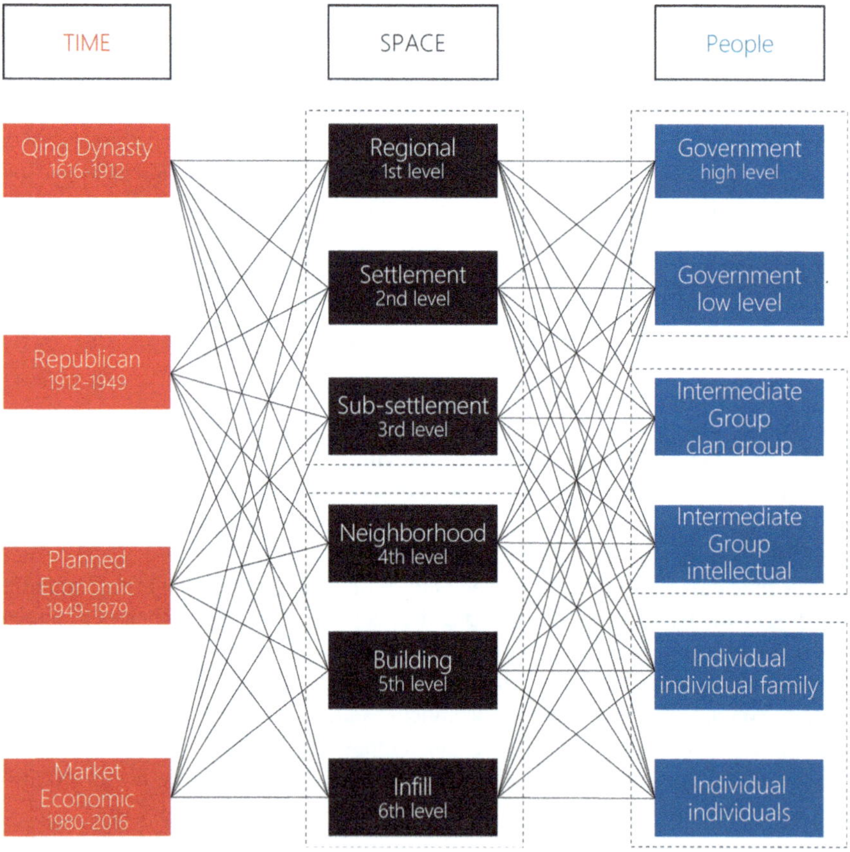

Fig. 5.1 Diagram of matrix of time, space, and people

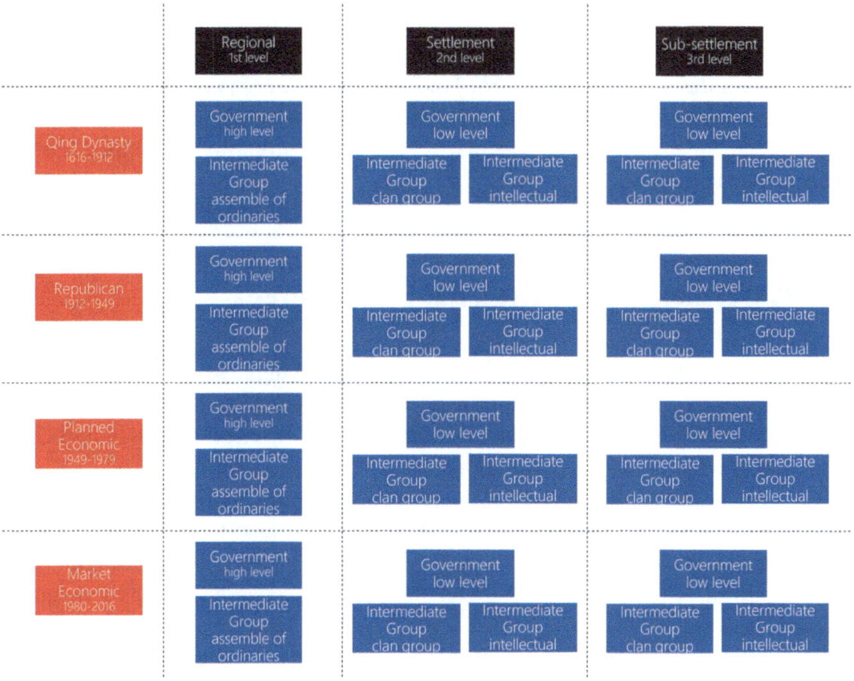

Fig. 5.2 Controlling agents of settlement levels in different periods

1. The Government (Official Agent): This group represents the state administration. It includes formal governmental bodies at various levels, such as the county, township, and village committees. Its power is derived from the legal and political system, and it regulates rural space through policies, laws, and official planning.
2. Intermediate Organizations (Semi-official/Social Agent): This group functions as a link between the state and the individual. In the pre-modern era, it primarily consisted of lineage groups (clans) and local intellectuals. In the modern era, entities like the production brigade served this role. These organizations manage collective resources, coordinate public affairs, and enforce social norms within the community.
3. The Individual (Folk Agent): This group comprises individual households and residents. They are the primary users of the built environment and make decisions regarding their private living spaces and daily activities.

The roles of agent groups vary depending on the construction phase and time period. The variation in the roles of agent groups forms an interlocked matrix of time, space, and people, reflecting the complex interactions among different agent groups during the construction of the built environment. The tangible and intangible environment, which includes societal, economic, political, cultural, and technological aspects, keeps evolving. This evolution has brought about simultaneous changes and generated a series of contradictions among these aspects.

This chapter focuses specifically on the roles of the Government and Intermediate Organizations. These two agents are the primary forces of "top-down" governance. They possess the authority to control land distribution, organize public infrastructure projects, and define the overall morphology of the settlement. The role of the Individual agent, which operates primarily through "bottom-up" adaptation, will be examined in Volume II.

5.2 Governance at the Regional Level (Level 1)

Regional level is the first level of the built environment in the seven levels set in this study. The observations at the regional level are made from a macroscopic perspective on a broad scale. The observations at the regional level reflect how human beings compromise with a broader setting of nature and society.

The agent group that can control the transformation at the regional level is of a large-scale nature, showing how the joint forces of human beings in the specific context affect the general trend of social transformation. The human entities at this level consist of two groups, namely, the high-level government and the masses (Table 5.1).

High-level government refers to the official units that take responsibility for the management of the specific region discussed in the research. It is a top-down administrative system that is always constituted of a series of governmental departments with different levels from the central authority to the local governments whose sphere of jurisdiction is not smaller than the discussed region. For example, the control agents that take responsibility for the whole area of Taishun County should be government departments with a level not lower than the corresponding county. In this case, when considering Xiaqiao Village in the context of the entire Taishun County region, the responsible government agents contain the Taishun County government, the Wenzhou City government, the Zhejiang provincial government, and even the central government of the country.

The central government can influence the regional built environment by releasing new policies. For example, in order to suppress the rebellions caused by local bandits,

Table 5.1 Controlling agent of the regional level

	Official agent		Intermediate agent			Individual agent	
	Government (high level)	Government (low level)	Assemble of ordinaries	Lineage	Intellectuals	Family	Individuals
Qing	√		√				
Republican	√		√				
Planning	√		√				
Market	√		√				

new counties containing Taishun County were established in the mountainous area of Zhejiang and Fujian provinces during the middle of the Ming dynasty. This decision made by the Ming court directly strengthened and normalized the management of the mountainous area, and fostered the regional development greatly. The policy impact from the government has been strengthened, especially in recent years, as a series of policies of the New Village Plan have been released successively, such as the combination of adjacent villages or the moving of peripheral villages. It has changed the rural spatial pattern at the regional level greatly.

The power exerted by the official government on the village is demonstrated in a more general way. After all, the change of policies is the absolute power that completely transforms society. For the Xiaqiao case, four releases of new decrees impact the transformation of the regional pattern. The first event was the establishment of Taishun County during the Jingtai period of the Ming Dynasty. The official administration of this area attracts more people to immigrate and settle here, which enlarges the scale of the settlement. For Xiaqiao, three of the four lineages now living here have immigrated after the official establishment of the county.

Secondly, the democratization of ancestral halls was allowed by Daliyi and subsequent court orders. It was a radical change for all rural society. After that, ancestral halls became an indispensable element, always treated as the spiritual center in clan villages. The last two policies were the militarization and de-militarization of Taishun County. During the transition from the Ming to the Qing Dynasty, Taishun had been an area controlled by the Southern Ming court, and the whole of Taishun had been a battleground scrambled by both forces. After that, in the Kangxi period, due to defending against the rebellion of Sanfan (three generals, 三藩之乱), military features were strengthened. Several fortresses were set up, one in Xiaqiao. The ancestor of the Tang lineage immigrated here as a soldier at that time. But after society was restored to stability, the military settings were canceled by the government. Xiaqiao returned to a common village again.

The other control agent that impacts the regional built environment in rural China is the force from the assembly of ordinaries. It refers to the living modes, customs, and culture. These aspects reflect the preferences and habits shared by a specific group of people and have been handed down through generations. It is the cultural gene of the locals that leaves an imprint on the result of the regional spatial pattern. For example, the daily life circle of the villagers is formed based on the topographical conditions under a certain transportation capacity. Meanwhile, it shows a tendency of preference according to local traditions. For example, the marriages of the villagers in Xiaqiao are mainly concentrated within a regional circle with a radius of 10 km, due to traffic conditions of daily activities and the availability of communication with similar dialects. But, in all probability, some localities are more popular than others, so more marriages are formed between there and Xiaqiao. It depends on the closeness between the villages, or sometimes even between the two lineages.

Generally, at the first regional level, the transformations are controlled by superior power from nature, the national government, or the whole society, which are beyond individual control. They follow the general trend of social development. But environmental coherence is not fully attributable to the presence of hierarchical structures of form and place [1]. Thus, the regional level is the foundation of the remaining five levels, and all the settings and transformations that occur at this level can be seen as the background and premise for the following levels.

5.3 Governance at the Settlement Level (Level 2)

The settlement level is the second level in the study of settlement patterns. It is influenced by the regional level and inherits certain features from it, forming a hierarchical relationship in the study of settlement development. The settlement level reflects the way in which a rural community is formed, organized, and grows.

The major agent that directly controls the transformation of housing settlements at this level has changed. During pre-modern China, it was local intellectual groups and lineage groups. After the establishment of the PRC, it became the government (Table 5.2). It generally can be divided into three stages: the self-organized stage, the transitioning stage, and the cogoverned stage.

5.3.1 Self-organized Stage: Qing

The self-organized stage covers the periods of the Qing dynasty and the Republican period in this study, when the official power was not sufficiently strong to directly govern villages. In pre-modern China, the lowest administrative level in the formal administrative system approved by the court was the county government. It was responsible for the entire county, including hundreds of villages. Given the limited administrative capacity of the county government in pre-modern times, especially considering the complexity and trivial nature of these affairs, it could not participate in the daily affairs of a single village. Instead, the direct connection between the government and the village concerned tax collection, which was important to the government. Occasionally, the government dealt with lawsuits and resolved serious disputes that could not be coordinated and settled by the villagers themselves.

Table 5.2 Controlling agent of the settlement level

	Official agent		Intermediate agent			Individual agent	
	Government (high level)	Government (low level)	Assemble of ordinaries	Lineage	Intellectuals	Family	Individuals
Qing		√		√	√		
Republican		√		√	√		
Planning		√		√	√		
Market		√		√	√		

In the Qing dynasty, the governing capability of the official force was weaker. As a result, the responsibility was instead assumed by the folk organizations, which were the direct governing agents of the village from the perspective of settlement. The folk organizations that controlled rural society and directly affected the transformation of settlement patterns consisted of two groups: clans and intellectuals. Conceptually, these two groups were independent of each other, but in real-life situations, there were instances of overlap.

5.3.1.1 Clan Group and Intellectual Group, Direct Agent of Control on Settlement Level

The clan or lineage, in the context of pre-modern China, is a social group formed by several families based on blood relationships. It is organized according to the order of blood relationships. It is a cluster of dozens of families connected by kinship ties. In the context of Chinese rural society, which is established and functions based on a network of blood relationships, the clan is the basic organization for governing the village. It maintains daily operations.

A village can be divided into one or several clan groups based on different blood relations. The clan system within a group plays a role in governance and is responsible for coordinating the affairs between different families. Then, different clan groups compete and coordinate their interests.

The clan system has three approaches in governing rural society, namely, clan genealogy, ancestral hall, and clan regulations. Clan genealogy is used to identify members who belong to the same clan group. It also serves as a crucial means of connecting the entire clan. The ancestral hall, where clan activities are held, also symbolizes authority in the village. Clan regulations are the norms of clan-related behavior, and clan activities regulate the rights and obligations of each family member.

Lineage is an important power to acquire and control land in imperial China. For example, in imperial China, most lands in Guangdong province were controlled by powerful lineages, which formed many powerful lineages with significant governance influence that could not be ignored. In Xiaqiao, the power of lineages is not as strong as that in Guangdong, but its influence on land governance cannot be ignored. The way lineages control land is through lineage property, which includes lineage farmland, lineage homesteads, and lineage mountains. Lineage is a united consortium of families. Theoretically, families belonging to one lineage should have blood connections. Since the blood connection cannot be easily identified, some families with the same surname unite as a lineage union to compete with other lineages for local governance. Lineages always build their own ancestral halls and have a certain amount of land to cover the costs of lineage activities, such as the maintenance of the ancestral hall, ancestral worship rituals, and so on.

In addition, land transactions were conducted by the villagers themselves. All land transactions should be under government supervision. However, given the time and monetary costs of obtaining government endorsement, most land transactions are conducted privately. The villagers involved in the transaction always find witnesses to formulate the deed and supervise the implementation of the transaction.

The intellectual group is another intermediate folk organization that helps rule rural society. Intellectuals hold absolute authority in rural society due to the importance and scarcity of knowledge. In pre-modern China, most common villagers were illiterate. However, many necessary transactions could only be completed with written language. Thus, the intellectuals, who held the power of discourse, were assigned by the government, according to its set standards, to represent it in dealing with folk affairs.

The local intellectual group is the intermediary between the government and the common villagers. It is responsible for conveying government policies to every family. Policies cannot be directly conveyed from the government to the villagers. A third party is required to execute the orders. Therefore, the government influences the transformation of settlements by releasing related policies. However, the direct operators of these policies are the lineages and the intellectuals.

The county government always endorses the revered elderly or respected intellectuals as its representatives in the village to assist in settlement management. The inscribed boards, made by county officials, are presented to honor the respected elders. Most affairs within the lineage were resolved among themselves instead of being reported to the government.

The control of the intellectual group can even extend across villages and lineages. This group can organize construction activities and regional rituals through the joint efforts of several villages. In the Xiaqiao case, the constructions of the Xidong bridge, the Linshui temple, and the Chen Daweng temple were organized by local intellectuals from neighboring villages and finally completed with the combined forces of all the involved villagers.

5.3.1.2 Controlling Approaches

These two groups, working together, become the direct force that governs the village and can transform the settlement pattern of the village. There are mainly two approaches that the two groups can contribute to the formation of settlement patterns.

The first approach is to build public spaces. The public buildings or other public constructions such as bridges are constructed by the joint forces of all the involved lineages under the leadership of one or two respected intellectuals as initiators. The construction of temples, bridges, and roads is all organized under this model.

The second approach is managing, controlling, and coordinating the behaviors related to the construction of common villagers' dwellings. In land transactions, they usually rely on the lineage system. Daily transactions of lands are operated by the villagers themselves. They are usually invited as witnesses and guarantors of the transaction to help draft and record the contract. When there are disputes over the ownership of land, the lineage and respected elites need to coordinate with the two parties concerned. Besides, the construction of an ancestral hall is fully controlled by the lineages themselves. There are some principles for organizing the space within the ancestral hall and its surrounding areas. It has a great impact on the spatial pattern of the whole settlement.

The common villagers are the real builders of the settlement, in fact. The common individuals group together as family units to build dwellings and cultivate farmlands. The formation of the settlement is closely related to every common villager. After all, they are the builders and users of the settlement. Although the construction behaviors are regulated by government-set official standards and folk organizations, they can still change the settlement layout by building new houses or reclaiming new lands. From this perspective, the common individuals, grouping together in the form of a family, are involved in the process of the formation of the spatial pattern of the settlement layout to some extent. Thus, during the Qing dynasty, the Chinese rural society was organized by the clan and intellectual groups.

5.3.2 *Transitioning Stage: Republican China*

The rural society in the Republican period has inherited features from the previous system of the Qing dynasty. However, the power from the government and lineage is shifting, with the former increasing while the latter decreasing.

5.3.2.1 Shift of Power Structure

In the early twentieth century, in the central and local political arenas, the regime changed dramatically. The government gained more power to control villages directly; it encroached upon the rural areas of autonomy. All central and regional regimes attempted to deepen state power to the grassroots level, regardless of their purpose. They believed that extending state power in this new way was the most effective means to control rural society.

In this process, the state power relied on the middle broker to plunder the resources. These brokers form their own forces in the village and squeeze away the original rural gentry elite groups, undermining the original local autonomy structure [2–4].

Regime reform is not only in the bureaucracy but also in the administrative system at the county level, so that it can more effectively implement the national policy. In the early years of the Republic, the "district" was an autonomous unit, but in 1915, after the constitutional reform of Yuan Shikai,[1] the "district" became the lowest

[1] Yuan Shikai 袁世凯 (1859–1916): He is an important military strategist, politician, diplomat and leader of the Northern New Army at the end of the Qing Dynasty. He had made important contributions to the modernization of modern China, including politics, economy, military, education and other aspects of the restoration of traditional Chinese culture. He is the the founder of modernized Chinese police, of China's judicature. He played a key role in the establishment of the Republic of China. He also is the former Prime Minister of the Chancellor of the Qing Dynasty, the Provisional President of the Republic of China and the President of the Republic of China.

level of government organization, at least in theory, directly controlled by the county government.

The "district" was established between 1908 and 1914. It is a high administrative level beyond the village. The district offices were often located in villages and towns that were home to large market towns. These offices are the largest achievements of the state power in the early years of the Republican period. The records of Taishun County Archive, in the 18th year of Republican China (AD 1929), state that the entire Taishun County is divided into four districts. Xiqiao belongs to the fourth district, Sixi[2] [5].

Along with the expanded power of government, the original power of regional informal cooperation, whether from folk intellectuals or the lineage system, has gradually disappeared and been replaced by new village-level administrative institutions.

At the beginning of the twentieth century, the county government designated village deputies, who were responsible for maintaining new schools, building roads, and engaging in various public utilities for "modernization." As mentioned earlier, the "local groups" can monopolize the link between the state and the village, although their social status is not high. When the state regime confirmed that the intermediary between the rural communities was the staff of the district office and the village leaders, the original intermediate authority that consisted of the intellectuals and clans was greatly weakened.

After the 20th year of the Republican period (1931 A.D.), the National Government formulated a law and policy that sought to maintain a clear affiliation between all rural communities and the government. The national government implemented a "township" autonomy law. "Town" became the highest self-government organization, which was given the power and responsibility to complete higher-level decrees, levy taxes, and even apportion payments and submit taxes to higher levels.

5.3.2.2 Self-organization Mode Inherited from the Qing Dynasty

The twentieth century was a period of rapid change in the state and society of China. Family organizations and their status in the state and social governance underwent major changes. The entire society fell into an unstable situation of continuous war

[2] According to Taishun Xianzhi (County Annals of the Taishun County, 泰顺县志), in the early of the republican period, the administrative zoning of the county keeps the same with that of in the Qing Dynasty. In 1935, the county is divided into four districts. In 1938, the fourth district is changed to the Sixi District whose center is set in the Baifenqiang village.

and chaos. Meanwhile, new policies were released one by one, and the administrative sectors changed ceaselessly.[3] In this situation, the local clan groups basically maintained the daily operation of the village.

In the first half of the twentieth century, clan organizations still existed in various parts of the country, and some even maintained a fairly complete form of organization, including genealogy, family hall, clan structure, family units, and patriarchal authority. In many cases, the state still relied on the family to manage the society. According to the investigation of Yixu in Fujian province by Lin Yaohua between 1933 and 1934, the clan organizations still had strong cohesion and dominated the governance of rural grassroots society. The government continued to rely on the family to exercise power and control practices [6].

However, since the beginning of the twentieth century, family organizations disintegrated in many ways. The relationship between family and state power also fissured. The impact of the war was partly due to catastrophic events and the migration of peasants, and, more importantly, the collapse of the political feudal monarchy. The economic commodity and the economic invasion of the Western countries caused the bankruptcy of the villages and the differentiation and impoverishment of the peasants. The spread of new culture, the empowerment of individuals, and the critique of the feudal family system and its ethics shook the economic, political, legal, and cultural foundations upon which clan survival depended. The traditional family organization was difficult to operate.

During the decline of family authority and organizational relaxation, family organizations and official ties were tense and interrupted. The tense and interrupted ties between family organizations and official sectors were the most prominent manifestation of the shaking of the traditional gentry status and the decline of the squire class itself. In history, the squire is often the representative of the rural society and the family, and also the intermediary between the rural society and the state. The squire played a leading role in the organization and management of the rural society and the connection between the rural family and the government. At the beginning of the twentieth century, the status of the gentry was greatly affected, and the decline in the countryside caused the overall decline in the gentry status. The abolition of the imperial examination system cut off the official future of the gentry. The rise of Western learning shook the gentry from their dominant status. This condition greatly reduced the legitimacy, cohesion, and authority of rural organizations, but also hindered the normal transfer of state power to rural society.

[3] According to the Taishun Xianzhi (County Annals of the Taishun County, 泰順縣志): During the Republic of China, the Northern Expeditionary Army troops into the Fujian provice. Then the defeated soldiers Zhou Yinren 周蔭人 have attacked Taishun. At the beginning of the new PRC period, Du Mouping 杜謀平 and Wu Zonglun 吳宗論 with their army also attacked Taishun, and ravaged the village.

5.3.3 *Co-governance Stage: Planned Economic and Market Economic PRC*

After the establishment of the People's Republic of China (PRC), the administrative efficiency improved greatly, causing the official power to go directly into the field of a single village; thus, the original self-organized field was encroached upon by the locals. The government gradually assumed the leading role in village management through several steps, such as land control, building control, and comprehensive planning control.

5.3.3.1 Enhancing Governance from Government

Land Reform

Land control is the first step for the government to participate in the construction and development of the village. Controlling the constructions within the village is the prerequisite for the government to participate in the village's building environment, given the basic role of land in the construction process. It changes the private ownership of land into collective ownership, leading to a series of interrelated changes from the perspective of construction.

The land policy has significantly changed since the imperial period. Immediately 40 years after the establishment of the PRC, the land policy in rural society turned over drastically. It changed from private ownership to primary cooperative, to senior cooperative with collective ownership, to people community, and eventually to the household contract responsibility system [7].

The formal reform of land policy started from the release of the Outline Land Law of China in 1947. It established equal land privatization in rural society, that is, "the ownership of land belongs to peasants, and all the lands were redistributed according to the actual population in the village, regardless of sex and age" [8]. The redistribution of land returned to a more balanced situation of land control in the village, similar to a combination of the lands of Xiaqiao. Previously, most lands were occupied by a few rich families. However, after this period, all people obtained a field of land with balanced size and quality.

After the 1950s, the contradiction between small peasant families and large-scale agricultural production was solved. The cooperative movement of agriculture gradually enhanced the transformation of land ownership from private to collective.

Primary Cooperation

Since the release of the Resolution on mutual cooperation in agricultural production in December 1951, the primary cooperation was established between rural families. This type of organization encouraged peasants to join the cooperative by becoming land-sharing members. However, the organization still maintained private land ownership [8].

Superior Cooperation

In 1953, the primary cooperation transitioned into superior cooperation, which marked the transition from private land ownership to collective land ownership. The transition was fully completed in 1956. In the new superior cooperative, the ownership of land was controlled collectively.

From 1958, the combination of small cooperatives into the people's commune on a larger scale became the mainstream. Several townships were merged together into a bigger township with six to seven thousand families. Along with the combination of townships, the lands belonging to these townships were combined to plan and farmed as a whole. In 1962, the three-level hierarchy of the people's commune included the people's commune, the production brigade, and the production cooperative. The blind coalition of lands was stopped. The land unit returned to the production cooperative. It was the basis of the land policies in the following decades.

From the 1970s, household production replaced collective production. Although land ownership did not change, it endowed peasants more rights to claim their residual interests. Ownership and management rights shifted again. Peasant families had their own homesteads and a certain amount of farmland. Although these lands still belonged to the collective, the peasants had their own usage rights. After the series of land policies, the government currently completely controls all the lands in the village. It can change the use of land according to specific regulations. However, it must negotiate with the residents who own the usage rights of the land. The two approaches to acquire a homestead include inheriting from the older generation or getting approval for a new piece of land from the government. The second approach is almost impossible at present due to the strict limitation of the quota for new homesteads.

Building Control

Apart from the land, the construction of individual buildings is also regulated by the government. In the planned-economy period, the construction of houses was limited by the distribution of materials. In the supply system, all the resources were distributed in a unified manner. Strict quotas of building materials for every production brigade existed. Thus, the buildings were built only with the materials supplied and approved by the government.

After the economic reform in the 1980s, building materials can be bought freely from the market. The houses can be built in any form and style according to the preference of the owner. However, the government has set some regulations to control the height, form, roof, and facade. These regulations are released strictly by the government. To maintain the traditional appearance of the village, buildings cannot be higher than six floors around the central area of Xiaqiao. The facade that faces the main street should be traditionally decorated. To unify the visual sense, the government organized the project to renovate the facade and rooftop of the buildings along the main tourism street.

Planning Control

The spatial pattern and land function are adjusted by the government through planning, which includes multi-level planning, such as the planning of Wenzhou City, the planning of Taishun County, and the planning of Sixi Town. The planning with lower levels should fulfill the regulations of the planning with higher levels.

The planning that directly influenced the spatial pattern of Xiaqiao was the overall planning of Sixi Town and the development planning of the Langqiao Cultural Park. The former targeted the central area of Sixi Town that includes two villages, namely, Xiaqiao and Baifenqiang. It adjusted the usage of land, construction area, and spatial pattern from an overall perspective. The latter emphasized the future development of the Cultural Park in Xiaqiao. It planned the different zones with different functions in the park, and differentiated the suitable and unsuitable areas for construction. It further designed the landscape of the park.

The construction activities in Xiaqiao should follow the guidance in the two plans. It is the new approach that the government uses to participate in the construction process.

5.3.3.2 Disappearing Governance from Lineages

The clan and local intellectual groups, as the intermediate agents that controlled the rural society in the Qing dynasty, disappeared after the establishment of the PRC. In the great change of ideology, the local intellectuals were always mistaken for landlords who were identified as enemies in the class struggle. The clan system was identified as a remnant that hindered social improvement of the pre-modern society. These two groups were the rural elite classes who were responsible for self-governance in the rural society. However, both intermediate groups of folk organizations were prohibited by the government in the planned-economy period due to the collectivization ideology that was predominant across the country. The self-organization system collapsed in rural society with the disappearance of the folk elite class in this society.

The governance power at the basic level was transferred to the production brigade in the planned-economy period. During the collectivization period, Xiaqiao was reorganized as Xiaiqao production brigade. It became the governance unit at the grassroots level to organize production activities, distribute resources, and deal with daily affairs.

After the economic reform in the 1980s, the production brigade system disintegrated. It was replaced by the village committee. Today, the village committee is a special administrative organization within the Chinese official system. It is the basic unit belonging to the official system, but is also an autonomous organization that is elected by the villagers themselves. The village committee of Xiaqiao includes representatives from all five different lineages in the village to negotiate the distribution of rights and responsibilities.

After the 1980s, the clan system was revived. However, its role in the everyday life of rural society changed. The government no longer directly managed the village. Instead, it turned into a cultural symbol and became a cultural phenomenon. At present, the five lineages in Xiaqiao have routine rituals to worship their ancestors every year. They have rebuilt their ancestral hall and rewritten the family archives. However, all these activities are different from actual life and production. The lineage system no longer controls the production resources and participates in public affairs. It is only a family activity for spiritual assurance.

Of the four family surnames in Xiaqiao, two are derived from immigrants who arrived during the Song Dynasty: the Jinan (济南) Lin arrived during the early North Song Dynasty, and the Chen arrived at a later period. The former group became the dominant power that controlled the whole village. Most of the Lins lived in settlements, including Houchi (后池, after pool), Huayuan (花园, Garden) and Xi'an (西安, safely west). Their power even expanded to other adjacent areas as their offspring multiplied. The Chens were not strong enough to compete against the Lins, and they were forced to live in a peripheral area of the village named Nanshandi (南山底, Under Nanshan) until the late Qing Dynasty. The other two surnames, Xihe (西河) Lin and Tang lineage, are derived from immigrants who arrived at the end of the Ming Dynasty and the early Qing Dynasty, respectively. A co-administrative system was formed among the four groups of descendants, and this strongly influenced the development of the village.

The settlement of Xiaqiao has grown over time. For information on rural society and the corresponding changes in settlement and architectural forms, see Table 5.3 and Table 5.4.

1. Due to the increasing population, more land was chosen for house-building. Although the density of the village is still low, it is slowly increasing. The two periods of rapid construction correspond to the two periods of rapid increases in population.
2. There have been various reasons behind the village's expansion. During the agricultural period, in response to the pressure brought by population growth, new dwellings were built on the slopes of the mountain or on the open spaces between previous dwellings to protect farming lands. During the post-agricultural period, new dwellings were hurriedly built alongside the new road, and so former farming land was converted into building sites.
3. After an unstable period marked by wars and disturbances, Chinese society entered a more peaceful era. Agricultural productivity was enhanced, and new techniques were introduced. For example, during Phase II, a new food crop, sweet potato, was introduced. This higher-yield crop could feed more people,

Table 5.3 Changes in the structure of rural society across different stages of development

Phase No.	Period	Political system	Social situation	Economic system/ industry type	Population/ family structure	Local govern power	Technological innovation	Social changes
Phase I	Qing Dynasty Phase I	Premodern society (Qing Dynasty)	Instable, wars and disturbance	Premodern/ extremely rely on agriculture, periodical fair	Low level/ large family	Force of lineage	N/A	N/A
Phase II	Qing Dynasty Phase II	Premodern society (Qing Dynasty)	Stable, few disturbance	Premodern/ extremely rely on agriculture, periodical fair and few fixed shops	Great increasing/ large family	Force of lineage	Well-developed wood structure, new crop introduced	Slowly changed/ more flourish than before
Phase III	Republic Phase	Republic China	Instable, wars and disturbance	Transitional/ less rely on agriculture, (periodical fair, more fixed shops, regional trade)	Slowly increasing/ from large family to main family	Force of lineage	New material bricks	Slowly changed / become a hub of transportation during the war to fight Japan
Phase IV	PRC Phase I	Socialist state	Rather stable	Planned economy/ less rely on agriculture, (periodical fair, more fixed shops), few industry	Slowly increasing/ main family	Government	Bricks, concrete	Slowly changed / establishment of road
Phase V	PRC Phase II	Socialist state	Stable	Market economy/ less agriculture, more commercial, and other service industry (travelling), with many sops	Great increasing/ nucleus family with few peoples	Government, and the renaissance of lineage force	Bricks, concrete, and much advanced techniques	Changed greatly/ more establishments of road

Table 5.4 Changes in settlement and architectural forms across different stages of development

Phase No.	Period	New settlements in village	Location	Major developed lineage	Number of newly constructed dwellings	Type of newly constructed dwellings	Public buildings I	Public buildings II
Phase I	Qing Dynasty Phase I	Houchi Huayuan Xi'an Nanshandi	Mountain coves, the foot of mountain	Houchi branch of Jinan Lin	Few	Single dwelling, small courtyard dwellings (few)	Beijian Bridge	Ancestral hall of Houchi Lin
Phase II	Qing Dynasty Phase II	Houzhuanyang, Yangtou, Dalingjiao, Nanshan, Zhaidi, Daligang, Beishanqiao	From Mountain cove to slope, from the foot of mountain to plain ahead	Cooperation of all the four lineages	Many	Large courtyard dwellings, single dwellings	Rebuilt Beijian Bridge, Linshui Temple, Chendaweng Temple	Ancestral hall of Xihe Lin Li, Ancestral hall of Tang
Phase III	Republic Phase	Xidong	Areas surrounding the old dwellings	Cooperation of all the four lineages	Few	Single dwellings with three to five Lius	N/A	N/A
Phase IV	PRC Phase I	Area along the road	Along the road	Cooperation of all the four lineages	Moderate	Single dwellings with three Lius	Great hall of Xiaqiao, primary school	N/A
Phase V	PRC Phase II	Area along the road	Along the road	Cooperation of all the four lineages	Many	Rowed single dwellings with one Liu	Middle school, public pavilions serving for travelling	Rebuild the ancestral hall of Xihe Lin, a new ancestral hall of Houchi Lin

and so the population increased. During Phase V, the village's stable social environment was characterized by greatly enhanced productivity, a rapid increase in population, and a great accumulation of personal wealth. In recent years, more advanced building techniques have been introduced, resulting in large-scale housing construction. During Phase II and Phase V, there was effective management of public spaces. In Phase II, a public temple was built beside Dongxi Bridge (东溪, East River), a trade center was raised beside Beijian Bridge (北涧, North River), and several ancestral halls served as residential centers.

4. Overall, the settlement has transformed from a dotted-distribution model (based on independent families) to a cluster-distribution model (expanding families) and then to a linear-distribution model (the result of modern 'nuclear' families).

5.4 Governance at the Sub-settlement Level (Level 3)

Sub-settlement Level is the third level that inherits the features from the second settlement level and is influenced by the first regional level. It addresses settlement issues at the smallest scale.

The situation of the controlling agent at the sub-settlement level is similar to that at the second settlement level. The mode of control changes from local self-government to government official planning. The major controlling agent transforms from intermediate folk organizations in pre-modern times into the government in modern times (Table 5.5).

Before the establishment of the PRC, the development of the sub-settlement was self-governed. However, compared to the joint force of multiple lineages, the sub-settlement is often controlled by one or two dominant lineages that constitute the main residents. Peasants often reside with their lineage groups, so a residential quarter, which functions as a sub-settlement, is often dominated by one or two major clans. The controlling force from the dominant lineage clearly determines the spatial pattern of the corresponding sub-settlement.

During the Qing dynasty, the government, lineage, and individuals had different roles in controlling rural society. The government exercised hierarchical control over land management. However, land was controlled by clans, which influenced

Table 5.5 Controlling agent of the sub-settlement level

	Official agent		Intermediate agent			Individual agent	
	Government (high level)	Government (low level)	Assemble of ordinaries	Lineage	Intellectuals	Family	Individuals
Qing		√		√	√		
Republican		√		√	√		
Planning		√		√	√		
Market		√		√	√		

the objects, means, and prices of transactions. Powerful lineages protected their properties.

After the government's power entered the village sphere in 1949, sub-settlements gradually came under direct official control. Compared to the settlement level, the government had a stronger influence on sub-settlements. This caused drastic changes due to the weaker resistance of a single sub-settlement. For example, the new residential area built by the government in the northern part of Huayuan, locally known as Daligang, completely changed the original spatial pattern. The new type of space and neatly arranged houses were inserted into the naturally developed rural built environment on a large scale. This could hardly be achieved by the villagers' self-management force alone and could only be done by the force of the government.

At the sub-settlement level, the targeting group, as the agent of control, was the same as that at the settlement level. However, its impact, through the forces from the intermediate group or the government, was more evident at the sub-settlement level. The controlling forces were more predominant in transforming a sub-settlement than the entire settlement.

The controlling agent groups at the three levels are different. From a spatial perspective, government power has a greater impact on higher-level entities than on lower-level ones. From a temporal perspective, government power has increased over time.

5.5 The Shift of Macro-control Power

The analysis of governance at the regional, settlement, and sub-settlement levels reveals a distinct historical shift in the locus of macro-control power.

In the pre-modern period (Qing Dynasty and Republican Era), the control of rural settlement space was primarily held by intermediate organizations, specifically lineage groups and local intellectuals. While the imperial government established the administrative boundaries of the county, it did not directly intervene in the internal spatial management of the village. Lineage groups managed land ownership, organized the construction of public infrastructure such as bridges and temples, and regulated the settlement's morphology through customary rules and Fengshui principles. This represented a system of local self-governance.

Following the establishment of the People's Republic of China in 1949, this power structure changed fundamentally. The state integrated the village into a centralized administrative system. During the planned-economy period, the production brigade replaced the lineage as the unit of resource management. In the subsequent market-economy period, although economic production shifted back to households, the government's control over spatial planning intensified. The local government now holds the legal authority to regulate land use, approve construction sites, and enforce environmental and tourism planning codes.

Consequently, the primary agent of macro-control has transitioned from the social authority of the lineage to the administrative authority of the state. The definition of settlement patterns and boundaries now depends more on formal government policies than on local customs.

5.6 The Limitations of Macro-control

Although the government and intermediate organizations establish the structural framework of the settlement, their control has inherent limitations. Macro-level governance effectively determines the allocation of land, the layout of road networks, and the location of public facilities. These decisions define the physical boundaries and legal constraints of the village.

However, these macro-agents do not fully dictate the specific spatial practices of individual residents. While the government can designate a plot of land for housing and regulate its height, it cannot precisely control the internal layout, the choice of non-structural materials, or the daily usage of the space by the family. There is a gap between the planned order established by top-down governance and the actual living environment created by residents.

Individual families and villagers operate within the framework set by macro-control, but they also adapt to it. They may modify building interiors to suit changing family structures or adjust construction details to fit economic budgets. These micro-level adaptations and construction behaviors represent a different logic of spatial transformation driven by "bottom-up" agency. This dynamic interaction between macro-constraints and individual action will be the focus of the analysis in Volume II.

References

1. Habraken N. J. and Jonathan Teicher (1998). The Structure of the Ordinary: Form and Control in the Built Environment. Cambridge, Mass.: MIT Pres. pp. 226.
2. Duara Prasenjit. (1988). Culture, Power, and the State : Rural North China, 1900–1942.
3. Wu Han 吴晗, Fei Xiaotong 费孝通. (1988). Huangquan yu Shenquan 皇权与绅权 (Emperor Power and Intelligence Power). Version 1. ed. Shehuishi Congshu 社会史丛书 (Collection of Social History).
4. Li Huaiyin 李怀印, Sui Yousheng 岁有生, Wang Shihao 王士皓. (2008). Huabei Cunzhi: Wanqing he Minguo Shiqi de Guojia yu Xiangcun 华北村治: 晚清和民国时期的国家与乡村 (Governing Villages: Country and Village in the Late Qing Dynasty and Republican China). Version 1. ed. Zhongguo Xiangcun Shehui Yanjiu Congshu 中国乡村社会研究丛书 (Collection of Chinese Rural Society). Beijing: Zhonghua Shuju.
5. Editorial Committee for County Annals of the Taishun County《泰顺县志》编纂委员会. (1998). Taishun Xianzhi泰 顺县志 (County Annals of the Taishun County). Hangzhou: Zhejiang People's Publishing House.
6. Lin, Y. H. (2013). The golden wing: A sociological study of Chinese familism. London: Routledge.
7. Zhou Zuwen周祖文. (2012). Zhongguo Nongcun Tudi Zhidu Bianqian: Yige Nongye Sheyu de Shijiao (1949–1985) 中国农村土地制度变迁: 一个农业剩余的视角 (1949–1985) (Land

Policy in Rural China: A Perspective of Agriculture). ed. Research of Modern History., Hangzhou: Zhejiang Daxue Chubanshe.

8. Chen Dan 陈丹, Tang Maohua 唐茂华. (2009). Zhongguo Nongcun Tudi Zhidu Bianqian 60 Nian Huimao yu Qianzhan 中国农村土地制度变迁 60 年回眸与前瞻 (Transformation and Future of the 60 Years of Chinese Rural Land Policy). In Chengshi 城市 (City), 2009.10: 41–45.

Chapter 6
Conclusion to Volume I

This chapter synthesizes the findings of Volume I regarding the macro-spatial transformation of Xiaqiao Village from 1644 to 2015. The analysis focused on the first three levels of the built environment: the region, the settlement, and the sub-settlement. These levels represent the large-scale physical and institutional structures that define the rural landscape. The research investigated how environmental conditions, lineage organizations, and formal government policies interacted over four centuries to establish the spatial framework of the village. The findings demonstrate a consistent shift in the forces driving spatial change. In the pre-modern period, the built environment was primarily shaped by geographical constraints and the social authority of lineage groups. In the contemporary period, these organic and local drivers have been replaced by centralized state administration and formal land-use planning. This transition has redefined the boundaries, connectivity, and internal organization of the village, moving it from a lineage-based natural settlement to an administratively regulated functional unit. Furthermore, this volume establishes that these macro-spatial levels function as the structural constraints—or the physical and legal context—within which rural life is organized. While the government and collective institutions now determine the overall morphology and land allocation of the settlement, they do not fully account for the specific living practices of individual households. By defining this institutional framework, Volume I provides the necessary foundation for examining the micro-level adaptations and individual agency that are the focus of Volume II.

X. Lin and H. Wang, *A Multi-Level Analysis of China's Rural Built Environment*, Urban Sustainability, https://doi.org/10.1007/978-981-95-4945-0_6

6.1 Summary of Macro-Spatial Transformations in the Settlement Levels

This volume has analyzed the transformation of the rural built environment in Xiaqiao Village by examining the first three levels of the seven-level analytical framework: the regional level (Level 1), the settlement level (Level 2), and the sub-settlement level (Level 3). These levels collectively define the macro-spatial framework and institutional context of the village.

At the regional level, the research demonstrates that the spatial development of the village is determined by the intersection of environmental constraints and administrative hierarchies. In the pre-modern era, the rugged topography and river systems of southern Zhejiang functioned as primary regulators, maintaining the village as a relatively isolated and self-sufficient spatial node. Over four centuries, this has shifted as the village became increasingly integrated into larger state-led infrastructure and economic networks. The transformation reflects a transition from a settlement governed by local geographical conditions to one defined by regional connectivity, such as the construction of provincial road networks and state-mandated administrative zoning.

At the regional level, the research demonstrates that the spatial development of the village is determined by the intersection of environmental constraints and administrative hierarchies. In the pre-modern era, the rugged topography and river systems of southern Zhejiang functioned as primary regulators, maintaining the village as a relatively isolated and self-sufficient spatial node. Over four centuries, this has shifted as the village became increasingly integrated into larger state-led infrastructure and economic networks. The transformation reflects a transition from a settlement governed by local geographical conditions to one defined by regional connectivity, such as the construction of provincial road networks and state-mandated administrative zoning.

At the sub-settlement level, the analysis reveals a fundamental shift in the internal organization of space, moving from lineage-based territories to functional zones. Historically, sub-settlements were formed as natural villages defined by kinship ties, where specific lineages (e.g., the Tang and Lin families) controlled distinct residential quarters. In the contemporary period, these social and spatial boundaries have blurred due to government-led relocation policies and land redistribution. Sub-settlements are no longer exclusively defined by lineage occupation but are now differentiated by their administrative and economic roles, such as designated commercial streets or protected cultural heritage sectors. This transition signifies the replacement of traditional social structures with formal planning mechanisms as the primary driver of sub-settlement transformation.

6.2 The Shift in Macro-governance Hierarchy

A central finding of this research is the identifiable shift in the governance hierarchy that regulates macro-spatial transformations. The authority to manage land, public space, and settlement morphology has transitioned from local social organizations to formal state administration. This evolution is analyzed across three distinct historical stages.

During the Qing Dynasty and the Republican period, the village functioned through a system of local self-governance. In this stage, the primary agents of control were intermediate social groups, specifically lineages and local intellectuals. These agents exercised social authority to manage the distribution of land and the allocation of homesteads within the community. They were responsible for the organization and financing of public infrastructure, such as the construction of bridges and temples, which served as the physical markers of lineage presence. Formal state power remained at the county level, intervening primarily in taxation and external security, while leaving the internal spatial order of the settlement to be determined by local customary rules and lineage negotiations.

Following the establishment of the People's Republic of China in 1949, the governance structure underwent a fundamental change. The state integrated the village into a centralized administrative system, effectively eliminating the previous self-governance model. During the planned-economy period, land ownership was converted from private to collective. The production brigade replaced the lineage as the primary unit of resource management and spatial control. Construction activities were strictly regulated through the centralized distribution of building materials and labor, ensuring that the physical development of the village served collective economic and political objectives.

In the market-economy period (1980–2015), the governance hierarchy evolved into a more complex regulatory system. While economic liberalization allowed individual households to manage their own production, the government's influence over spatial planning intensified through the application of modern administrative tools. The local government now exercises control through land-use regulations, the approval process for new homesteads, and the enforcement of planning codes related to environmental protection and tourism. Consequently, while the lineage system has seen a cultural revival, its role as a governing agent of physical space has been largely superseded. The authority to determine the overall form, functional zoning, and boundaries of the settlement now resides with the formal state administration.

6.3 From Structural Framework to Individual Adaptation

This volume has established that the macro-spatial levels—comprising the Region (Level 1), the Settlement (Level 2), and the Sub-settlement (Level 3)—function as the structural framework of the rural built environment. This framework is shaped by "top-down" forces, including the constraints of natural geography, the historical territories of lineage groups, and modern government land-use policies. Together, these forces determine the physical boundaries, the layout of infrastructure networks, and the legal parameters within which the village operates. The macro-spatial levels provide the institutional and physical context that maintains the overall stability and order of the rural settlement.

However, the establishment of this macro-level order represents only one dimension of the rural built environment. While the state government and collective institutions define the allocation of land and the general configuration of the village, they do not fully determine how these spaces are inhabited, utilized, or modified on a daily basis. The physical environment described in this volume serves as the "container" for rural life, establishing the pre-existing conditions for construction. Nevertheless, this macro-analysis does not account for the "content"—the specific spatial practices and construction behaviors of the residents. There remains a gap between the formal planning of the settlement and the actual living reality of the individual household.

To achieve a comprehensive understanding of the transformation of the built environment, it is necessary to examine how individual families and villagers operate within this established framework. The dynamics of housing construction, the adaptation of architectural forms, and the arrangement of interior spaces follow a logic of transformation driven by "bottom-up" agency. These processes are characterized by individual decision-making and incremental adjustments to the immediate living environment. These micro-level dynamics, encompassing the Neighborhood (Level 4), Building Cluster (Level 5), Individual Structure (Level 6), and Infill (Level 7) levels, are the subject of Volume II: Micro-Architectural Adaptation and Agency.

Appendixes

Appendix I: Changes in the Names of Townships in Taishun County

Appendix II The Transformation of Towns' Names inside Taishun
泰顺县内各镇、乡历史变迁

Town 镇名	**Town Name in Ming Dynasty**明代名称	**Town Name in Qing Dynasty** 清代名称	**Villages in the Town** 镇内所含村子
Luoyang 罗阳	The 59th Region of Yixiang Township, Rui'an county 瑞安县义翔乡五十九都		上交垟村、下交垟村、南外村
Sixi 泗溪	The 39th Region of Guiren Township, Pingyang county 归仁乡三十九都	The 6th District 六都	下桥村、白粉墙村、南溪村、前坪村
Baizhang 百丈	The 59th Region of Yixiang Township, Rui'an county 义翔乡五十九都	The 1st District 一都	百丈镇
Sankui 三魁	The 40th Region of Guiren Township, Pingyang county 归仁乡四十都	The 7th District 七都	薛宅村、张宅村、秀溪边村、庵前村、刘宅村、下溪坪村、水车垟村
Yayang 雅阳	The 38th Region of Guiren Township, Pingyang county 归仁乡三十八都	The 8th District 八都	新联村、塔头底村、百福岩村

(continued)

X. Lin and H. Wang, *A Multi-Level Analysis of China's Rural Built Environment*, Urban Sustainability, https://doi.org/10.1007/978-981-95-4945-0

(continued)

Shiyang 仕阳	The 40th Region of Guiren Township, Pingyang county 归仁乡四十都	The 6th District 六都	
Siqian 司前畲族	The 60th Region of Yixiang Township, Rui'an county 义翔乡六十都	The 1st District 一都	里光村、圆洲村、筱条村
Pengxi 彭溪	The 38th Region of Guiren Township, Pingyang county 归仁乡三十八都	The 8th District 八都	彭溪乡
Xiaocun 筱村	The 57th and 58th Region of Yixiang Township, Rui'an county 义翔乡五十七五十八都	The 4th District 四都	徐岙底村、柏村底村、东垟村、小岭村、前山村、东岙村、岩头村、坳头村、五蒲村、翁家山村、竹园村
Xiyang 西旸	The 58th Region of Yixiang Township, Rui'an county 义翔乡五十八都	The 6th and the 7th District 七都六都	
Guihu 龟湖	The 40th Region of Guiren Township, Pingyang county 归仁乡四十都	The 7th District 七都	龟湖镇
Xian'nian 仙捻乡	The 59th Region of Yixiang Township, Rui'an county 义翔五十九都	The north to the Luoyang city 城外北隅	仙居村、三滩村、下捻村、上捻村
Lingbei 岭北乡	The 59th Region of Yixiang Township, Rui'an county 义翔乡五十九都	The 1st District 一都	村尾村、上垟村、关际后村、北峰村
Nanyuan 南院乡	The 59th Region of Yixiang Township, Rui'an county 义翔乡五十九都	The 7th District 七都	
Zhuli 竹里畲族	The 59th Region of Yixiang Township, Rui'an county 义翔乡五十九都	The 1st District 一都	何宅垟村
Huangqiao 黄桥	The 59th Region of Yixiang Township, Rui'an county 义翔乡五十九都	The 1st District 一都	

(continued)

(continued)

Beipai 碑排	The 60th Region of Yixiang Township, Rui'an county 义翔乡六十都	The 1st District 一都	
Fengmen 峰门	The 56th Region of Yixiang Township, Rui'an county 义翔乡五十六都	The 3rd District 三都	
Lianyun 联云	The 57th Region of Yixiang Township, Rui'an county 义翔乡五十七都	The 3rd District 三都	箬阳村
Wengshan 翁山	The 57th Region of Yixiang Township, Rui'an county 义翔乡五十七都	The 3rd District 三都	
Fengyang 风垟	The 39th Region of Guiren Township, Pingyang county; The 57th and 58th Region of Yixiang Township, Rui'an county 归仁乡三十九都 / 义翔乡五十七五十八都	The 4th and 5th District 五都/四都	
Hengkeng 横坑	The 57th Region of Yixiang Township, Rui'an county 义翔乡五十七都	The 3rd District 三都	横坑乡
Fengwen 峰文	The 39th Region of Guiren Township, Pingyang county 归仁乡三十九都	The 6th District 六都	
Yuehu 月湖	The 38th Region of Guiren Township, Pingyang county 归仁乡三十八都	The 8th District 八都	
Xiahong 下洪	The 58th Region of Yixiang Township, Rui'an county 义翔乡五十八都	The 4th District 四都	上洪村、下洪村、贝谷村
Da'an 大安	The 58th Region of Yixiang Township, Rui'an county 义翔乡五十八都	The 7th District 七都	

(continued)

(continued)

Yangxi 垟溪	The 39th and 40th Region of Guiren Township, Pingyang county 归仁乡三十九都 四十都	The 6th and the 7th District 六都七都	
Zhouling 洲岭	The 39th Region of Guiren Township, Pingyang county 归仁乡三十九都	The 7th District 七都	
Wanpai 万排	The 40th Region of Guiren Township, Pingyang county 归仁乡四十都	The 6th District 六都	
Baoyang 包垟	The 56th Region of Yixiang Township, Rui'an county 义翔乡五十六都	The 3rd District 三都	
Xinpu 新浦	The 57th Region of Yixiang Township, Rui'an county 义翔乡五十七都	The 3rd District 三都	吴宅村、包宅村、库尾村
Jiufeng 九峰	The 39th Region of Guiren Township, Pingyang county 归仁乡三十九都	The 6th District 六都	
Songyang 松垟	The 39th Region of Guiren Township, Pingyang county 归仁乡三十九都	The 6th District 六都	
Liufeng 柳峰	The 38th Region of Guiren Township, Pingyang county 归仁乡三十八都	The 8th District 八都	
Xuexi 雪溪	The 40th Region of Guiren Township, Pingyang county 归仁乡四十都	The 7th District 七都	桥溪西村
Dongxi 东溪	The 38th Region of Guiren Township, Pingyang county 归仁乡三十八都	The 8th District 八都	
Hechao 鹤巢(无)	The 59th Region of Yixiang Township, Rui'an county 义翔乡五十九都	The east to the Luoyang city 城外东隅	

(continued)

(continued)

Nanshan 南山	The 59th Region of Yixiang Township, Rui'an county 义翔乡五十九都	The 1st District 一都	
Liguang 里光	The 60th Region of Yixiang Township, Rui'an county 义翔乡六十都	The 1st District 一都	
Zhuyang 竹垟	The 59th Region of Yixiang Township, Rui'an county 义翔乡五十九都	The 1st District 一都	
Huangkeng 黄坑	The 59th Region of Yixiang Township, Rui'an county 义翔乡五十九都	The 1st District 一都	
Yanshang 岩上	The 56th Region of Yixiang Township, Rui'an county 义翔乡五十六都	The 3rd District 三都	
Jujiang 吕江	The 56th Region of Yixiang Township, Rui'an county 义翔乡五十六都	The 2nd District 二都	
Hongkou 洪口	The 56th Region of Yixiang Township, Rui'an county 义翔乡五十六都	The 4th District 四都	
Yuxi 玉溪	The 57th and 58th Region of Yixiang Township, Rui'an county 义翔乡五十七五十八都	The 4th District 四都	
Geyang 葛垟	The 58th Region of Yixiang Township, Rui'an county 义翔乡五十八都	The 4th District 四都	
Xinshan 新山	The 57th Region of Yixiang Township, Rui'an county 义翔乡五十七都	The 3rd District 三都	
Nanpu 南浦	The 57th Region of Yixiang Township, Rui'an county 义翔乡五十七都	The 3rd District 三都	

(continued)

(continued)

Yuxi 玉西	The 39th Region of Guiren Township, Pingyang county 归仁乡三十九都	The 6th District 六都	
Nanxi 南溪	The 39th Region of Guiren Township, Pingyang county 归仁乡三十九都	The 6th District 六都	
Fuyang 富垟	The 38th Region of Guiren Township, Pingyang county 归仁乡三十八都	The 8th District 八都	
Wulipai 五里牌	The 38th Region of Guiren Township, Pingyang county 归仁乡三十八都	The 8th District 八都	
Xialu 夏芦	The 58th Region of Yixiang Township, Rui'an county 义翔乡五十八都/归仁乡三十九	The 4th and 5th District 四都/五都	
Zhanzhou 戳洲	The 39th Region of Guiren Township, Pingyang county 归仁乡三十九都	The 5th District 五都	
Xiujian 秀涧	The 39th Region of Guiren Township, Pingyang county 归仁乡三十九都	The 6th District 六都	
Dongling 东岭	The 38th and 39th Region of Guiren Township, Pingyang county 归仁乡三十八三十九都	The 6th and the 8th District 六都/八都	
Dong'an 东安	The 38th and 39th Region of Guiren Township, Pingyang county 归仁乡三十八三十九都	The 6th and the 8th District 六都/八都	
Jiku 积库	The 38th Region of Guiren Township, Pingyang county 归仁乡三十八都	The 8th District 八都	

(continued)

(continued)

Xuelin 雪临	The 38th Region of Guiren Township, Pingyang county 归仁乡三十八都	The 8th District 八都	
Zhangchen 章成	The 40th Region of Guiren Township, Pingyang county 归仁乡四十都	The 6th and the 7th District 六都七都	
Houzhang 后章	The 40th Region of Guiren Township, Pingyang county 归仁乡四十都	The 7th District 七都	
Zhangkeng 章坑	The 40th Region of Guiren Township, Pingyang county 归仁乡四十都	The 6th District 六都	

Appendix II: Marriage Circle of the Tang Family in the Xiaqiao Village

Village	Family	Distance	Region	Level	Times
Local (Benchu)	Chen, Jinan Lin, Xihe Lin	<500m	Local	The same village	26
Baifenqiang	Jinan Lin	<500 m	Neigbhoring Village	Administrative center in Sixi	9
Xikou		500 m–2 km	Neigbhoring Village	Village in Sixi	1
Banxi		500 m–2 km	Neigbhoring Village	Village in Sixi	2
Daling		500 m–2 km	Neigbhoring Village	Village in Sixi	1
Shangyuan		500 m–2 km	Neigbhoring Village	Village in Sixi	3
Sanchouyang		500 m–2 km	Neigbhoring Village	Village in Sixi	1
Zhengzhou	Chen	500 m–2 km	Neigbhoring Village	Village in Sixi	1
Shangyang		500 m–2 km	Neigbhoring Village	Village in Sixi	1
Nanxi		2–5 km	Neigbhoring Village	Village in Sixi	14

(continued)

(continued)

Village	Family	Distance	Region	Level	Times
Qianping			Nanxi Village		1
Shanhouzai			Nanxi Village		1
Bandiyang			Nanxi Village		1
Shangzhuang			Nanxi Village		1
Qishi					4
Yinshantang		2–5 km	Village	Village in Sixi	1
Sixitou		2–5 km	Village	Village in Sixi	4
Lishuyang		2–5 km	Village	Village in Sixi	8
Sanmenyang		2–5 km	Administrative Village	Village in Fengyang	12
Qiantou		2–5 km	Village	Village in Fengyang	5
Xiyuan		2–5 km	Village	Village in Sixi	14
Niuyangkeng		2–5 km	Village	Village in Sixi	2
Shendu		2–5 km	Village	Village in Sixi	1
Qianping		2–5 km	Yunyan Village	Village in Sixi	5
Xiaping		2–5 km	Yunyan Village	Village in Sixi	2
Shimen		2–5 km	Village	Village in Sixi	12
Qilong		2–5 km	Junzhunkeng Village	Village in Sixi	1
Sujiakeng		2–5 km	Village	Village in Sixi	1
Guoling		2–5 km	Village	Village in Liufeng	8
Zhuozhai		2–5 km	Village	Village in Liufeng	4
Hong'aoling		2–5 km	Village	Village in Liufeng	1
Zhujiaogang		2–5 km	Village	Village in Liufeng	1
Huangdanji		2–5 km	Village	Village in Liufeng	1
Xikou		2–5 km	Village	Village in Liufeng	1
Banlu		5–10 km	Village	Village in Fengyang	2
Wutongyang		5–10 km	Village	Village in Fengyang	4
Liyang		5–10 km	Village	Village in Fengyang	3

(continued)

(continued)

Village	Family	Distance	Region	Level	Times
Chongjiukeng		5–10 km	Village	Village in Fengyang	1
Luli		5–10 km	Village	Village in Fengyang	3
Xidi		5–10 km	Village	Village in Jiufeng	12
Shanchalan		5–10 km	Village	Village in Jiufeng	2
Aotou		5–10 km	Administrative Village	Village in Jiufeng	1
Xiujian		5–10 km	Village	Village in Yayang	4
Dong'an		5–10 km	Village	Village in Yayang	6
Liantou		5–10 km	Village	Village in Yayang	2
Shangrenyang		5–10 km	Village	Village in Songyang	1
Bankeng		5–10 km	Village	Village in Songyang	1
Dong'an		5–10 km	Village	Village in Yayang	6
Liantou		5–10 km	Village		2
Xiuxi		5–10 km	Village	Village in Dongxi	2
Dongxi		5–10 km	Administrative Village	Village in Dongxi	8
Liuzhai		5–10 km	Village	Village in Sankui	1
Wuyang		5–10 km	Village	Village in Sankui	1
Pengjiapu		5–10 km	Village	Village in Sankui	1
Jinzhuping		5–10 km	Village	Village in Sankui	1
Luli		5–10 km	Village	Village in Sankui	3
Wengshan		>10 km	Administrative Village	Village in Wengshan	12
Dishe		>10 km	Village	Village in Wengshan	1

(continued)

(continued)

Village	Family	Distance	Region	Level	Times
Hengkeng		>10 km	Village	Village in Hengkeng	1
Yantou		>10 km	Village	Village in Xiaocun	1
Kengxia		>10 km	Village	Village in Xiaocun	1
Xiuyang		>10 km	Village	Village in Yayang	5
Juxi		>10 km	Village	Village in Pingyang County	2
Luoyang		>10 km	Luoyang Town	Town Center	2

Appendix III: Chronology of Key Construction Agents in Tang Lineage

This table is synthesized and translated by the author from the Tangshi Zongpu (Lineage Archive). To focus on the relationship between lineage development and the built environment, only family branches and members directly involved in significant land transactions or construction activities are listed. Data regarding modern living descendants has been omitted for privacy.

Generation	Agent ID	Social Status (Education/ Rank)	Key Spatial Action (Construction/Land)	Period
G2	Tang Tingzuo	First Settler in Xiaqiao	Initial migration and settlement in Xiaqiao	1695–1762
G3-1	Tang Kaiming	/	Established social standing and residency in the village	1721–1785
G3-2	Tang Kaiwang	/	Contributed to initial family resource accumulation	1725–1776

(continued)

(continued)

Generation	Agent ID	Social Status (Education/ Rank)	Key Spatial Action (Construction/Land)	Period
G3-3	Tang Kayou	**Key Agent.** Received the "Highly Respected" social honor from the County Magistrate. Married into the wealthy Zhang family of Qianping	1794 (Qianlong 59): Initiated the construction of the large courtyard house at Houchi (Old Tang House). Established the family's permanent land rights and elite spatial presence	1730–1800
G4-1	Tang Chaoyu	**Honored Elder (En-qi).** Awarded an honorary plaque for family virtue by Prefect Lu Jinghua in 1829	**Key Action**: Initiated the construction of a new residential building at Dalingjiao (Houchi) during the early 19th century	1756–1837
G4-2	Tang Chaoliang	**Honored Elder (En-qi).** Received official recognition from Prefect Shen for exemplary moral conduct	Strengthened family social capital through marital alliances with the Chen lineage of Nanshan and the Zhong lineage of Fuding	1763–1833
G4-3	Tang Chaohan	Head of the third sub-branch of the lineage	**Key Action:** Organized the construction of a residential complex on the east bank of Houchi (Xiangyan area) in 1794	Late 18th Century
G4-4	Tang Wei	**County Scholar (Yi-xiangsheng)**. Achieved 3rd rank in the civil examination of 1794	**Key Action:** His academic success provided the institutional legitimacy and social status required for the family's transition to formal, high-standard courtyard architecture	1771–1821

(continued)

(continued)

Generation	Agent ID	Social Status (Education/ Rank)	Key Spatial Action (Construction/Land)	Period
G5-1	Tang Guoxiu	**Honorary Contributor Scholar (Li-gongsheng).** Recognized for high social standing	**Key Action:** Relocated to Yangtou in the early 19th century and initiated construction there. Actively organized the building of family shrines and the restoration of local bridges	1789–1879
G5-2	Tang Guoding	**National Academy Student (Guo-xuesheng)**	Maintained lineage land and social networks through marital alliances with the Lin (Houchi) and Wang (Guoling) familie	1791–1867
G5-3	Tang Zhao	National Academy Student (Guo-xuesheng)	**Key Action**: Initiated the construction of a residential complex in the Shanghu area of Houchi in 1868	1809–1868
G5-4	Tang Guangbo	Awarded an honorary plaque for family virtue by Prefect Lu Jinghua in 1829	**Key Action**: Initiated construction at the Dalingjiao area of Houchi. His social standing facilitated the family's spatial expansion in the early 19th century (Daoguang era):	Early–Mid 19th Century
G6	Tang Fazang	Married into the Zeng family (Sanmenyang branch)	Following an imperial decree, a stone memorial archway (Pai-fang) was erected at Xidong to honor his widow, Lady Zeng. This added a symbolic landmark to the village landscape in 1896	1837–1859

(continued)

(continued)

Generation	Agent ID	Social Status (Education/ Rank)	Key Spatial Action (Construction/Land)	Period
G7	Tang Gao	**County Scholar (Yi-xiangsheng)**. Achieved 6th rank in the civil examination of 1874	Expanded the primary residence by adding over ten liu (spatial units) including the left and right courtyard wings in 1899 and 1907. He also installed two circular gates at the entrance to regulate access	Late 19th Century

Appendix IV: Height, Structure, Material, Built Period, and Architectural Form in the Sub-settlement Level

The five architectural criteria show the increase of construction from the perspectives of quality and quantity.

Appendix IV-1

A: Transformation of Height: Beishanqiao

(continued)

(continued)

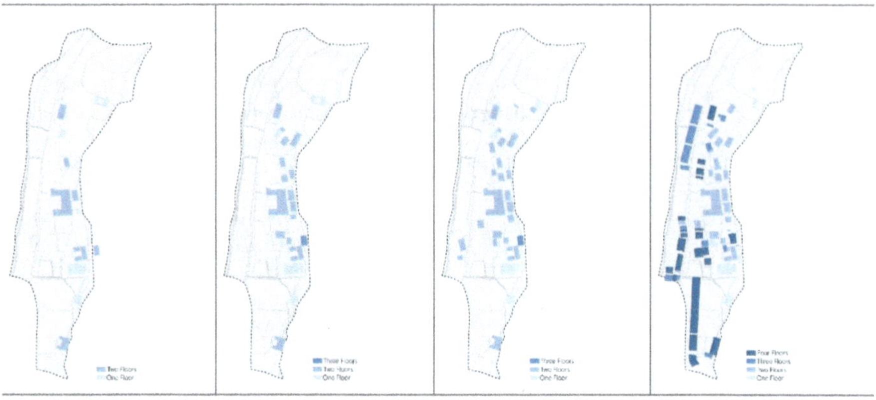

B: Transformation of Height: Houchi

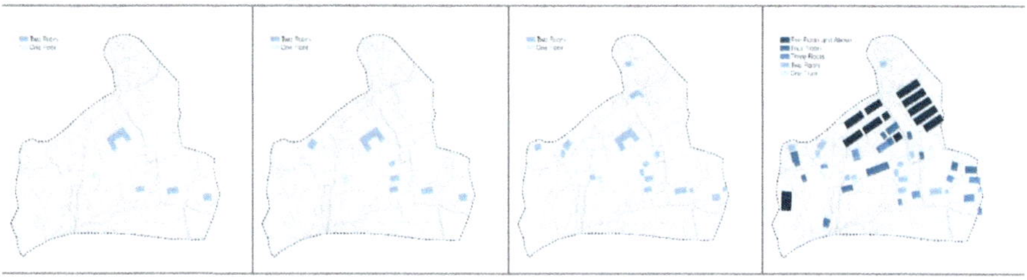

C: Transformation of Height: Huayuan

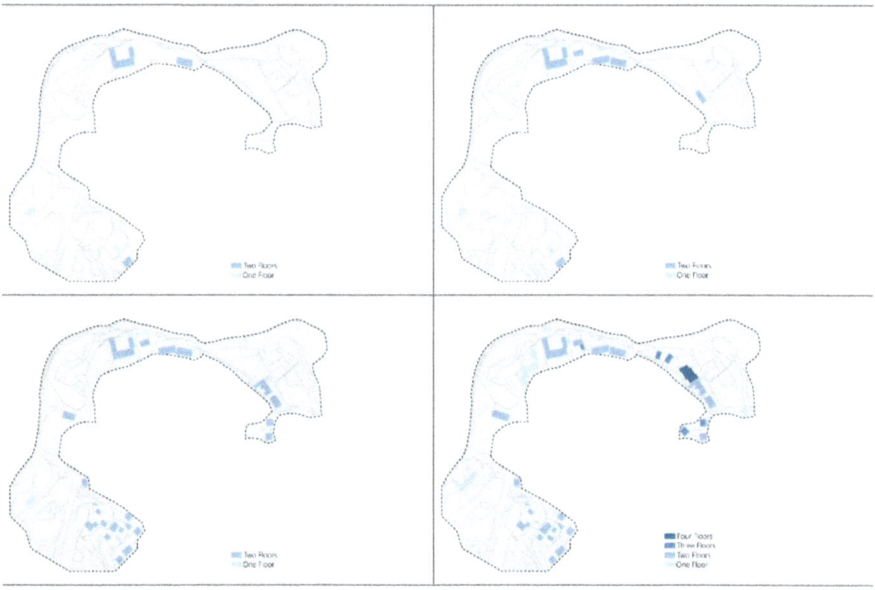

D: Transformation of Height: Nanshan

(continued)

(continued)

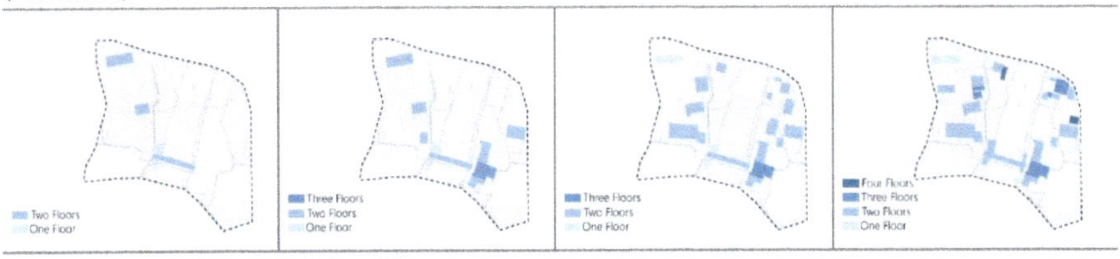

E: Transformation of Height: Qiaotou

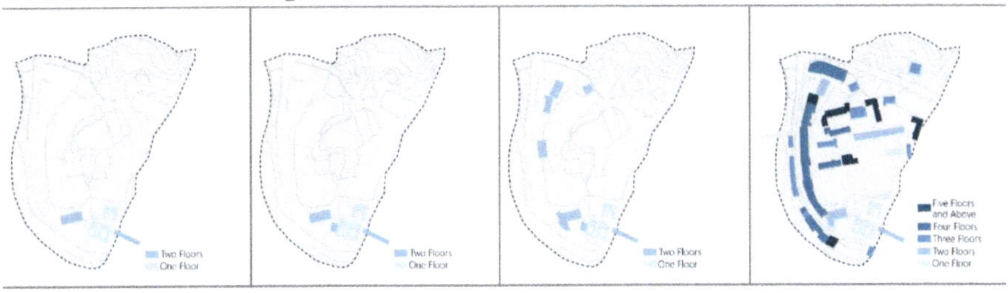

F: Transformation of Height: Xidong

G: Transformation of Height: Yangtou

Appendix IV-2

A: Transformation of Structure: Beishanqiao

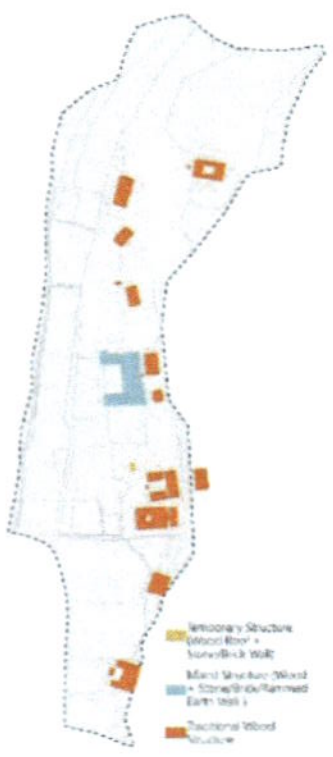

(continued)

(continued)

B: Transformation of Structure: Houchi

C: Transformation of Structure: Huayuan

(continued)

(continued)

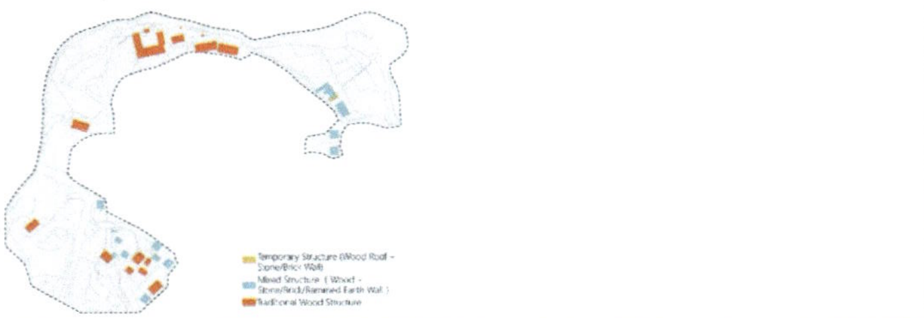

D: Transformation of Structure: Nanshan

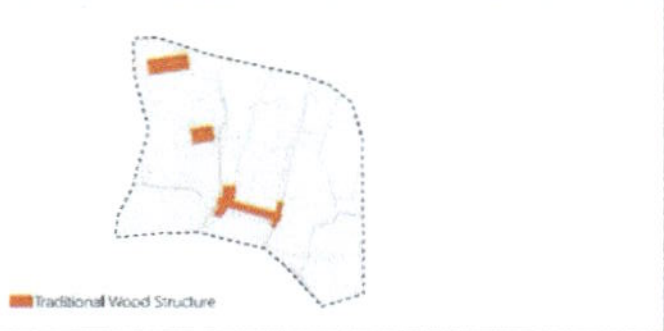

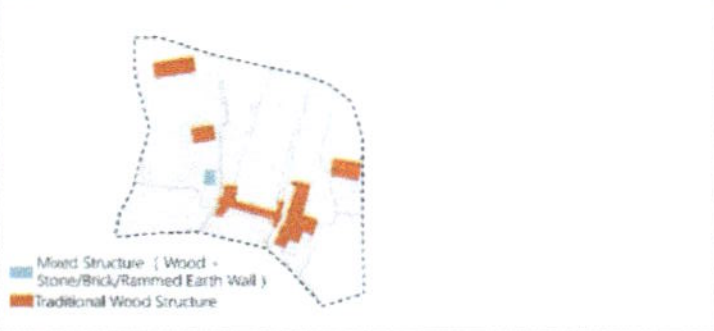

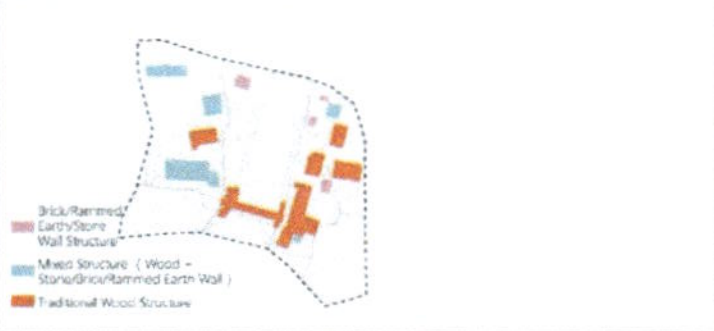

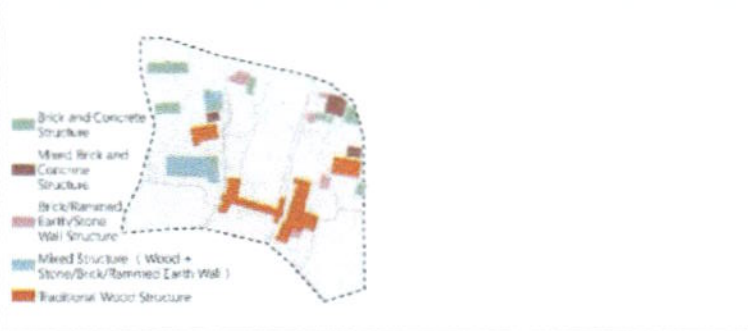

E: Transformation of Structure: Qiaotou

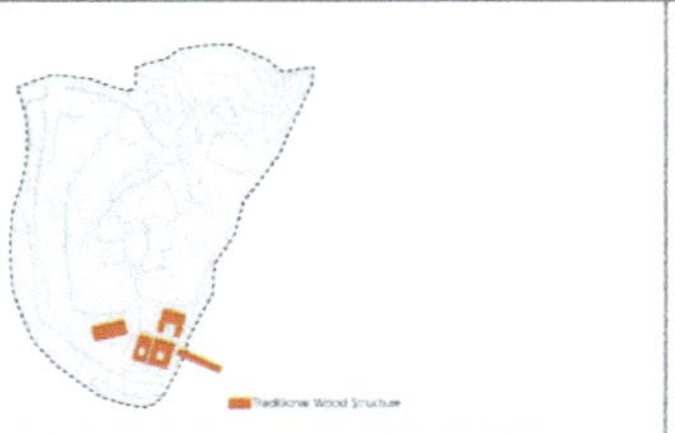

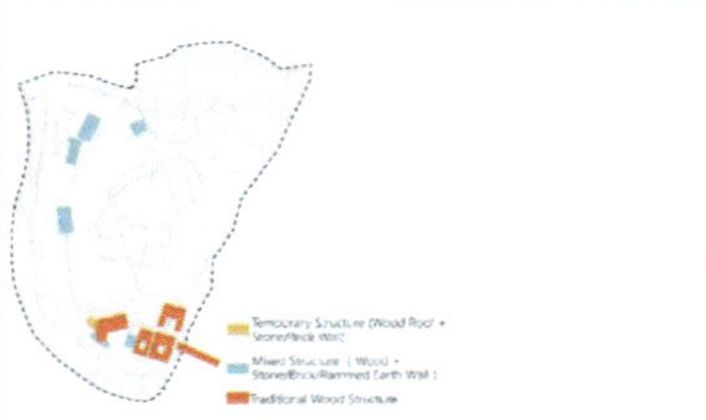

F: Transformation of Structure: Xidong

(continued)

(continued)

G: Transformation of Structure: Yangtou

Appendix IV-3

A: Transformation of Material: Beishanqiao

B: Transformation of Material: Houchi

(continued)

(continued)

C: Transformation of Material: Huayuan

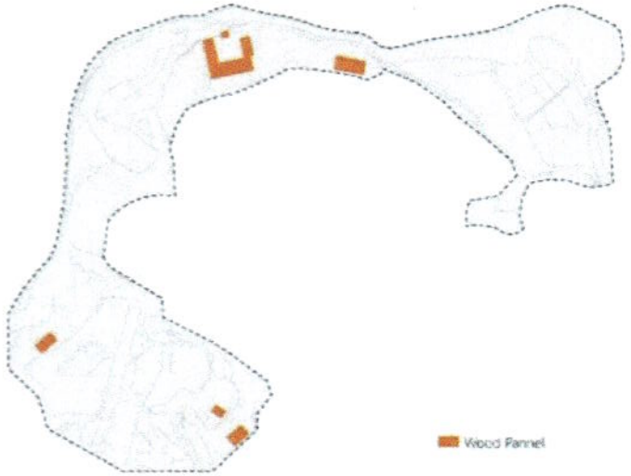

(continued)

(continued)

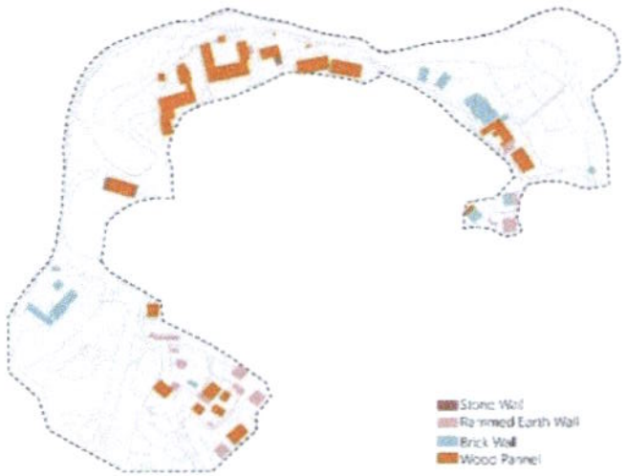

D: Transformation of Material: Nanshan

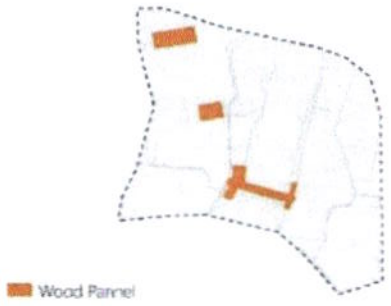

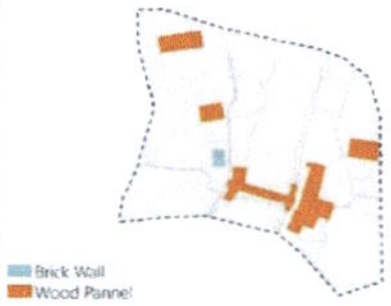

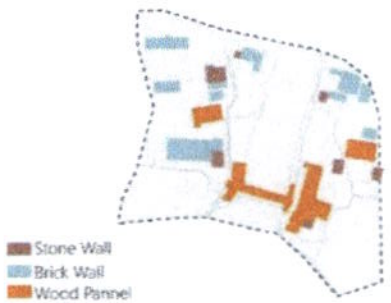

E: Transformation of Material: Qiaotou

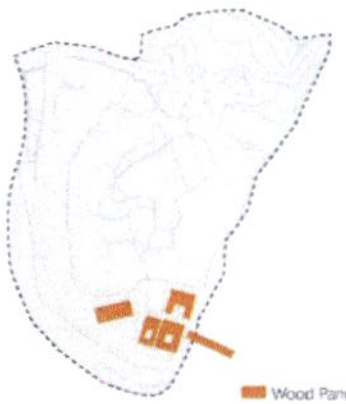

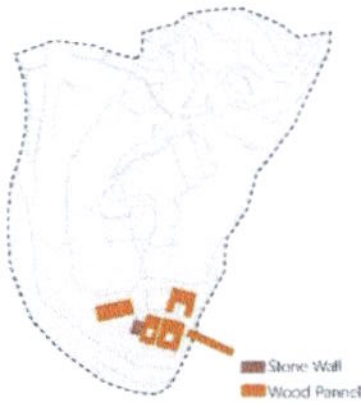

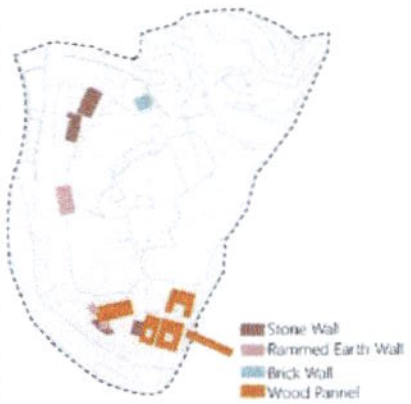

F: Transformation of Material: Xidong

(continued)

(continued)

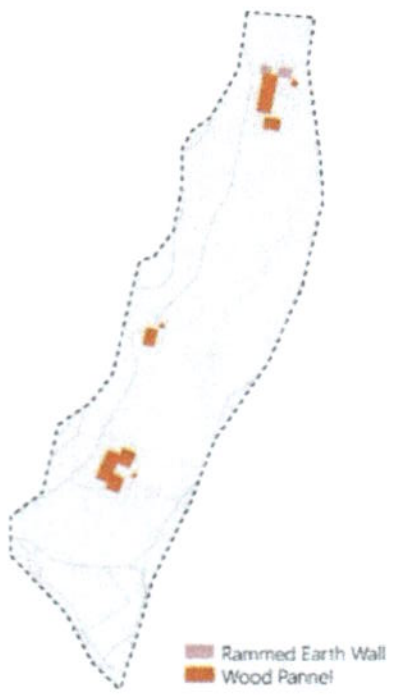	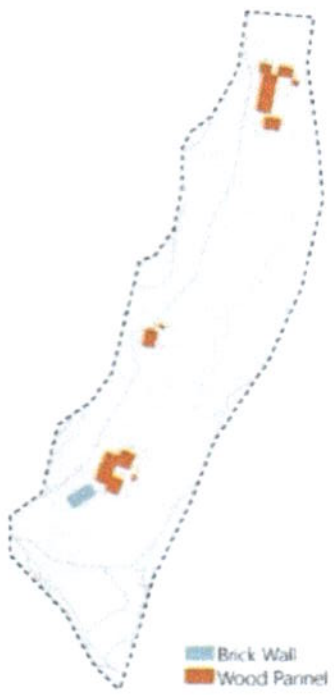		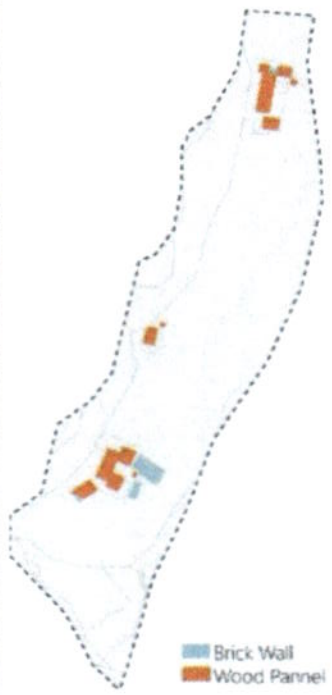

G: Transformation of Material: Yangtou

Appendix IV-4

A: Transformation of Period: Beishanqiao

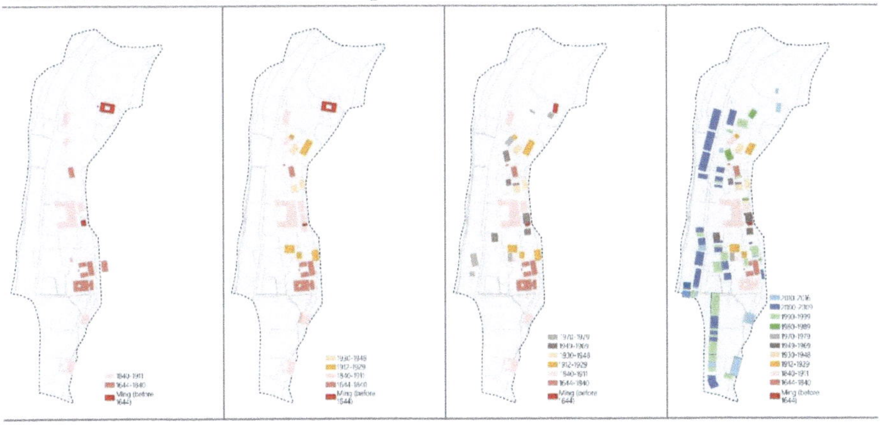

B: Transformation of Period: Houchi

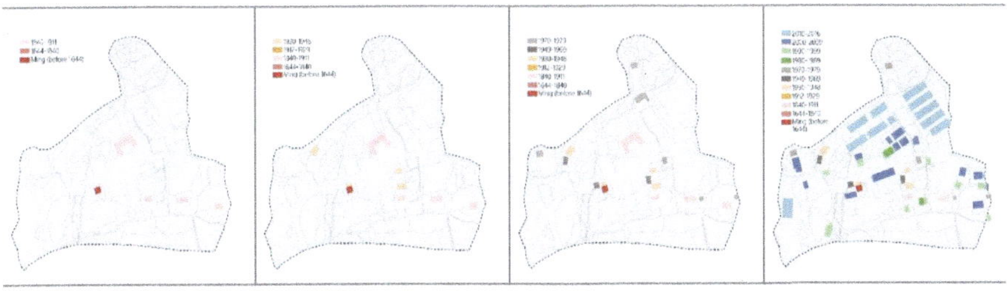

C: Transformation of Period: Huayuan

(continued)

(continued)

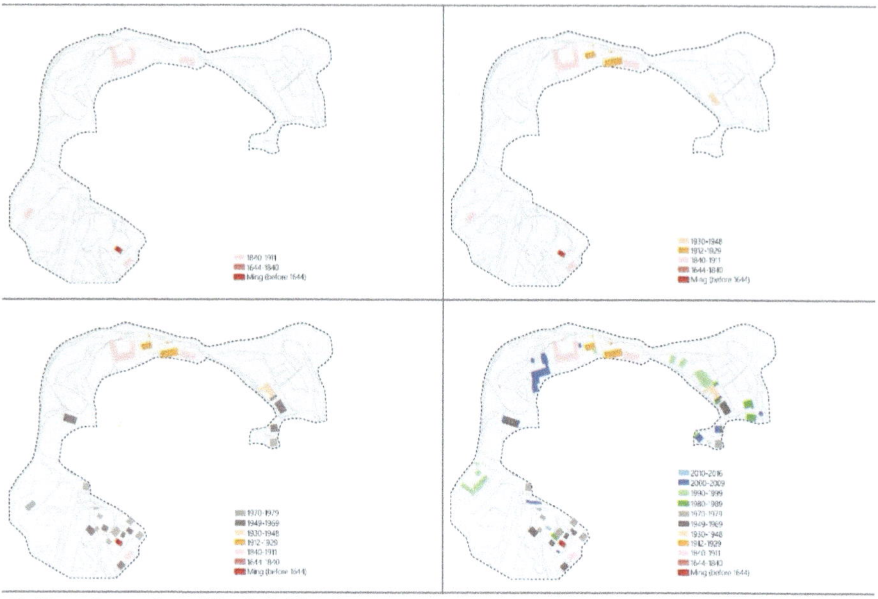

D: Transformation of Period: Nanshan

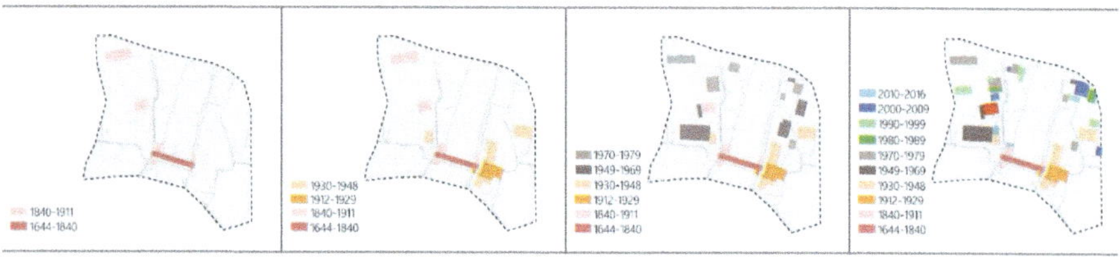

E: Transformation of Period: Qiaotou

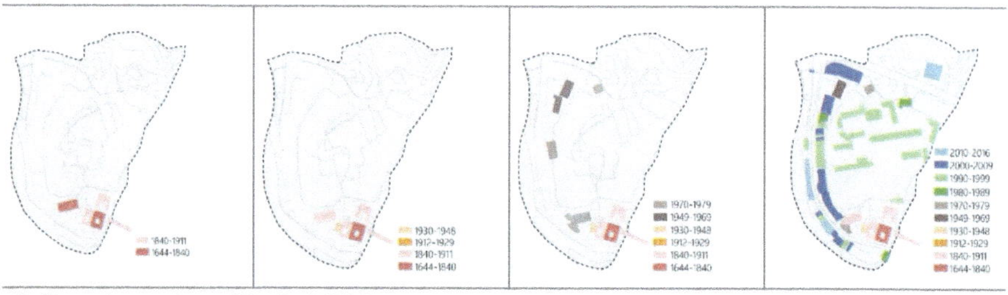

F: Transformation of Period: Xidong

(continued)

(continued)

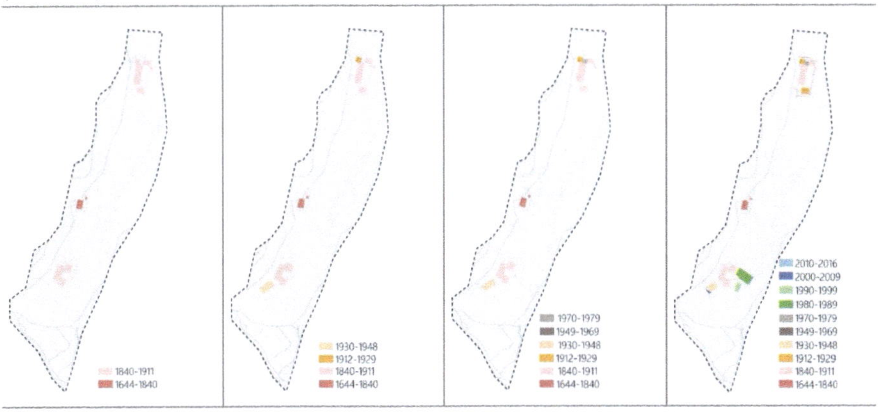

G: Transformation of Period: Yangtou

Appendix IV-5

A: Transformation of Form: Beishanqiao

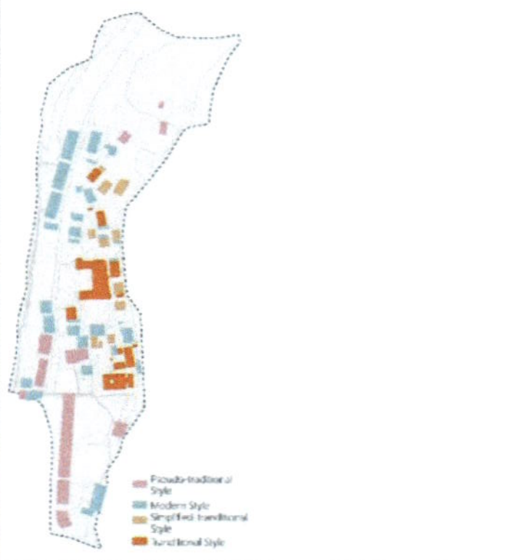

(continued)

(continued)

B: Transformation of Form: Houchi

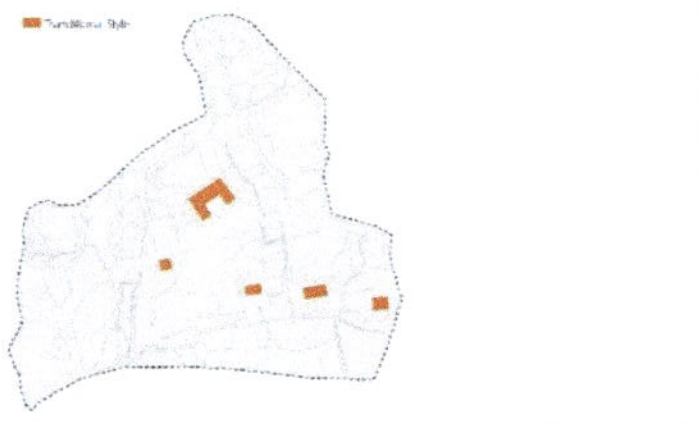

C: Transformation of Form: Huayuan

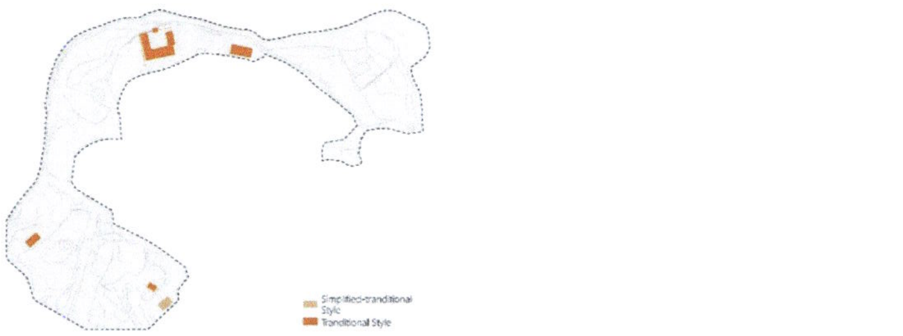

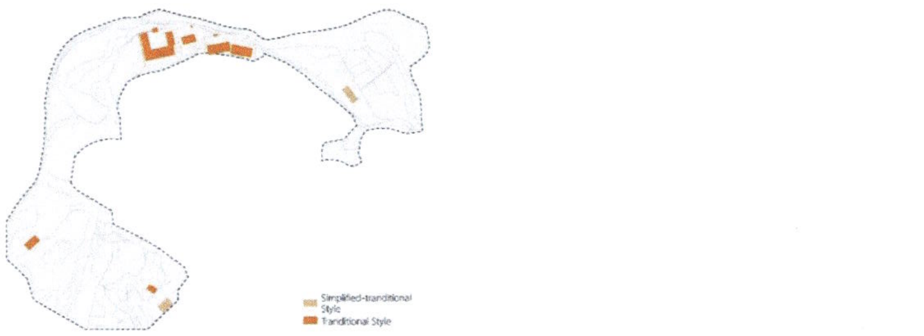

(continued)

(continued)

D: Transformation of Form: Nanshan

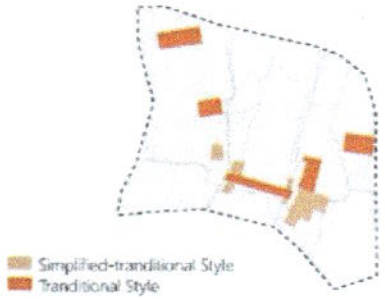

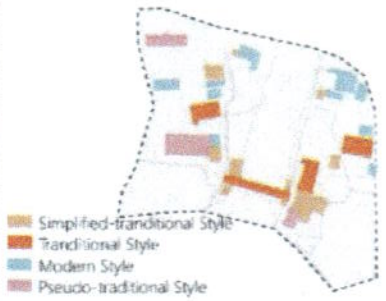

E: Transformation of Form: Qiaotou

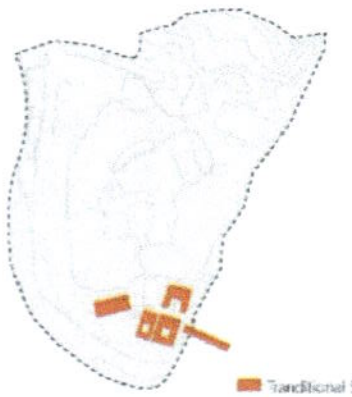

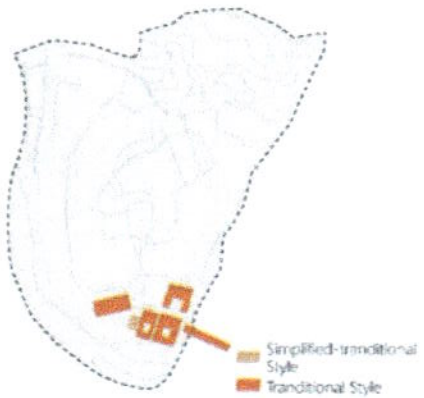

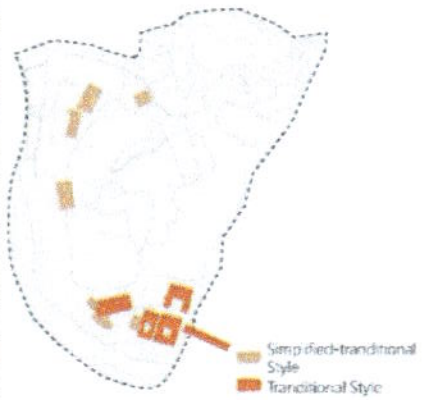

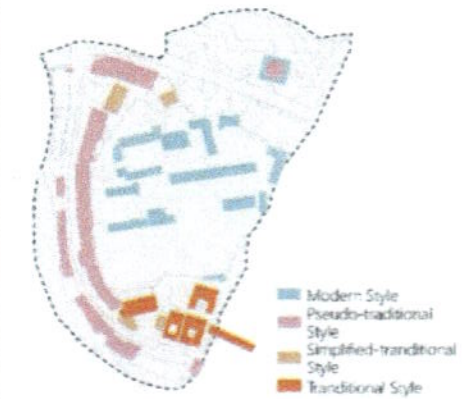

F: Transformation of Form: Xidong

(continued)

(continued)

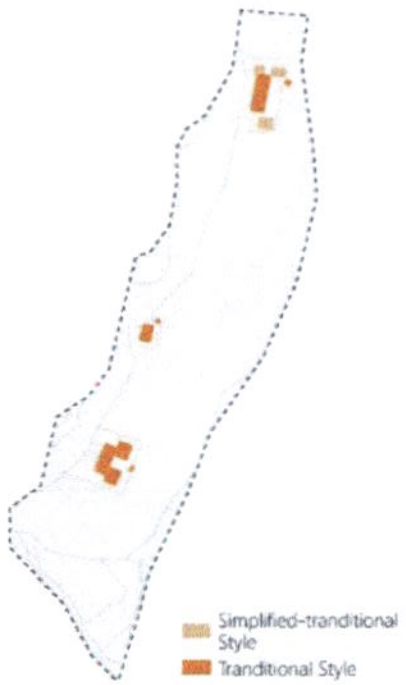

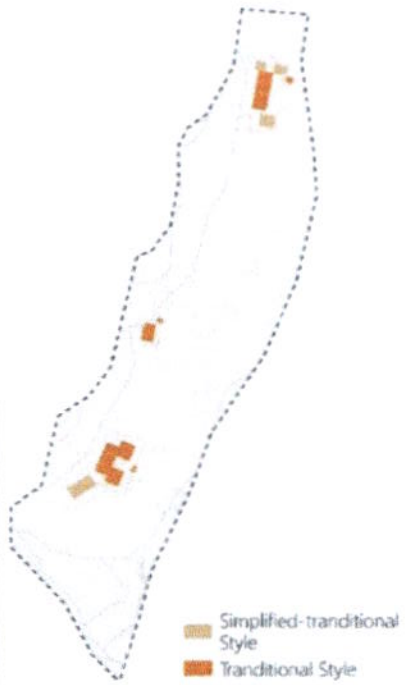

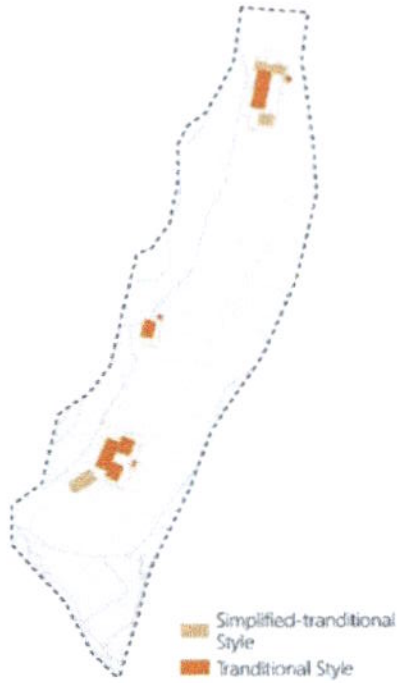

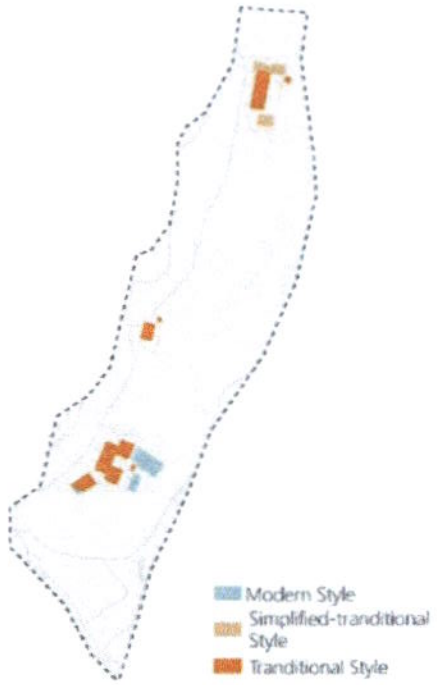

G: Transformation of Form: Yangtou

GPSR Compliance

The European Union's (EU) General Product Safety Regulation (GPSR) is a set of rules that requires consumer products to be safe and our obligations to ensure this.

If you have any concerns about our products, you can contact us on ProductSafety@springernature.com

In case Publisher is established outside the EU, the EU authorized representative is:

Springer Nature Customer Service Center GmbH
Europaplatz 3
69115 Heidelberg, Germany

Batch number: 10809441

Printed by Printforce, the Netherlands